Optimization
of
Transport Networks

Optimization
of
Transport Networks

Peter A. Steenbrink

A Wiley–Interscience Publication

JOHN WILEY & SONS

London · New York · Sydney · Toronto

629.04
S 814

Library of Congress Catalog card No 73-2793

ISBN 0 471 82098 9

Printed in Great Britain by
J. W. Arrowsmith Ltd., Bristol, England

Preface

Transportation is an important sector of society in which many decisions must be made. In the search for the right decisions to be taken, mathematical techniques for transport network optimization can be of great help.

In this book the transport network optimization problem is stated mathematically and then an extensive and detailed survey of the most important and best-known methods of solving this problem is given. For this survey special attention is directed to the applicability of the solution methods to large networks. Moreover a new technique is presented which yields a reasonable solution for large networks in a very short computation time. The application of this new technique to a very large existing network, namely the Dutch road network, is discussed as a case study in the second part of the book.

After the introduction of some general mathematical notions and the notations used in the book, the notions 'transportation' and 'transport network' are introduced. A transport network represents the real transport infrastructure used for transportion in the form of a number of nodes and links connecting these nodes, all having their own characteristics. A short treatment is given of the mathematical models describing the behaviour of travellers. The difference between a user-optimized and a society-optimized system is made clear: the former describing the situation resulting from (the simulation of) the real behaviour of all individual tripmakers, and the latter giving such traffic flows on the links of the network that the objective function for the society is optimized.

Next the transport network optimization problem is stated. Generally speaking in a transport network optimization problem, we investigate those values for the dimensions of the links (e.g. the numbers of lanes for the roads of a road network) and/or for the traffic flows on the links which optimize the objective function. Some possible and acceptable objectives are discussed, such as the minimization of the total social costs for a given trip-matrix and the maximization of the total social net benefits. Finally the fundamental constraints on the optimization are mentioned together with other possible constraints.

A full treatment is given of the best-known methods for the choice of the optimal dimensions of the links in a transport network. The dimension zero is included as a possibility, so that in fact a network structure is also chosen from a set of possible structures. The network optimization is a real combinatorial problem, so the use of combinatorial techniques, like branch and bound, to solve the problem is extensively dealt with. The size of the problem, however, is usually a great difficulty and, where this is so one will sometimes have to be content with a fairly good rather than an optimal solution. The use of heuristic techniques to get a fairly good solution can be very profitable but can also be very misleading. Examples are given of both features. Finally, aggregation and decomposition are discussed.

After surveying the main lines of the literature on network optimization such as are relevant to rather large networks, a new technique is presented for which only a very short computation time is needed. This time is of the same order as the computation time necessary for one assignment of the traffic to the network. The problem of costs minimization for a given trip-matrix is here decomposed into on the one hand a number of subproblems (yielding for every link the optimal dimension for any given traffic flow) and on the other hand a master problem in which the traffic flows are chosen in such a way that the objective function is minimized (and the constraints are met). This is accomplished by a stepwise assignment of the trip-matrix to the network according to the paths with the least value for the marginal objective function.

Next, some extensions of the problem are discussed such as the optimization over time, the optimization of networks for more travel modes and the optimal structure of a network.

Most of the computation time for network optimization is spent on finding the shortest paths in a network. It is, therefore, necessary to have very powerful shortest-path algorithms available. One chapter of the book is devoted to the most important and the fastest shortest-path algorithms, while special algorithms to recompute all shortest paths after small changes in the network are also discussed.

The second part of the book is devoted to a case study: the optimization of the Dutch road network. In this study, the method of the stepwise assignment according to the least marginal objective function is applied to get the optimal (minimum costs) road network for a given trip-matrix of car traffic. First the position of this costs minimization in the whole transport study is indicated. Next the definition of the objective function is given. Estimates are made for the different parts of the objective function such as travel time costs, vehicle-operating costs, costs of accidents, costs of investments and maintenance and costs to the environment. Relationships are defined between these costs and the dimensions of the links and the traffic flows on them. Minimization of these functions yields the optimal number of lanes in a road for any given traffic flow. The results of this are used in the master problem of defining the traffic flows in such a way that the total costs are minimized. The choice of the parameters of the stepwise assignment according to the least marginal objective function is treated in detail. Then the operation of the method is illustrated on some parts of the network and finally the results of the application of the method are compared with the results obtained by a heuristic procedure.

Putten, The Netherlands   PETER A. STEENBRINK
1973

Acknowledgements

This study has been carried out while working on the Dutch Integral Transportation Study, a study undertaken by the Netherlands Economic Institute at the behest of the Dutch Minister of Transport.

I am grateful to many people without whose help and encouragement this work would have never reached this final stage. In the first place I should like to thank Professor L. H. Klaassen of the Netherlands Economic Institute and the Netherlands School of Economics in Rotterdam and also Professor R. Timman of the Delft University of Technology for their much-valued advice and encouragement during all stages of the evolution of this work. Further, I am especially grateful to Dr. R. Hamerslag. He introduced me to the field of transportation research, directed my attention to the problem of transport network optimization and constantly proved a very valuable adviser and friend.

Many other people have contributed to this work. I would like to mention Professor J. Volmuller and J. A. Hartog and all workers on the Dutch Integral Transportation Study, particularly J. A. Bourdrez, B. Beukers, Dr. W. Horn and A. Rühl. Moreover, I am indebted to Mrs. H. A. Imhof, J. A. Kant, M. Koss and especially to J. A. Perton for coding the various computer programs and many other things.

The Directors of the Netherlands Railways and Dr. J. W. Geerlings must be thanked for providing me with the opportunity to write this book. Moreover I am grateful to the secretarial and drawing staff of the Netherlands Railways and the Netherlands Economic Institute for typing the manuscript and providing the figures and to Mrs. A. C. A. Elderson and the staff of John Wiley for their assistance in improving the style of the English.

Finally, the Dutch Minister of Transport must be thanked for giving me permission to use the material of the Dutch Integral Transportation Study for the second part of this book.

Putten, The Netherlands
1973
                                                    PETER A. STEENBRINK

Contents

PART I
THEORETICAL ASPECTS OF THE OPTIMIZATION OF TRANSPORT NETWORKS

PART I

Theoretical Aspects of the Optimization of Transport Networks

1

Some Mathematical Notions and Notations Used in this Book

In this chapter we will introduce briefly the basic mathematical concepts used throughout the book. In this way we also get the opportunity to point out the position of the problem and solution methods treated here in the whole class of related problems and solution methods. Finally, we will introduce the notation used.

1.1 OPTIMIZATION

1.1.1 The Definition of an Optimization Problem

In all fields decisions have constantly to be taken. Where the situation they deal with is very complex and, because of the impact they will have, it is important that the right decisions be made, mathematical optimization techniques can be a great help. For the purpose of this book the general mathematical optimization problem can be formulated as follows. It is desired to determine values for n variables $x_1, \ldots, x_n$ in such a way that the value of a function of these variables ($F(x_1, \ldots, x_n)$) is as large or as small as possible. The variables $x_1, \ldots, x_n$ are called *decision variables* or instrument variables. The function $F(x_1, \ldots, x_n)$ is called the *objective function*. The objective function must be *maximized* or *minimized*.

Moreover, there (may) exist certain relationships between the decision variables and/or the decision variables or functions of the decision variables must satisfy some inequalities or equalities. These are called the *constraints*:

$$g_h(x_1, \ldots, x_n) = 0 \qquad h = 1, \ldots, m_1$$

$$g_i(x_1, \ldots, x_n) < 0 \qquad i = m_1 + 1, \ldots, m_2$$

$$g_j(x_1, \ldots, x_n) \leqslant 0 \qquad j = m_2 + 1, \ldots, m$$

3

In this book we will write these general constraints as (using also the 'greater than' sign, which will occur if $g_i(x_1, \ldots, x_n) < 0$ is changed into $-g_i(x_1, \ldots, x_n) > 0$)

$$g_i(x_1, \ldots, x_n) <, =, > 0; \quad i = 1, \ldots, m$$

or as

$$\mathbf{G(X)} <, =, > \mathbf{0}$$

in which $\mathbf{X}$ is a vector with elements $x_1, \ldots, x_n$, $\mathbf{0}$ a vector for which all elements equal 0 and $\mathbf{G(X)}$ is a set of functions $g_1, \ldots, g_m$. We will always use bold capital letters to indicate vectors.

So a minimization problem is formulated as follows:

$$\left. \begin{array}{l} \min_{x_1, \ldots, x_n} F(x_1, \ldots, x_n) \\ \text{subject to } g_i(x_1, \ldots, x_n) <, =, > 0; \quad i = 1, \cdots, m \end{array} \right\} \quad (1.1.1)$$

also written sometimes as

$$\min_{x_i} F(x_1, \ldots, x_n)$$

$$\text{subject to } g_i(x_1, \ldots, x_n) <, =, > 0; \quad i = 1, \ldots, m$$

or in vector notation:

$$\left. \begin{array}{l} \min_{\mathbf{X}} F(\mathbf{X}) \\ \text{subject to } \mathbf{G(X)} <, =, > \mathbf{0} \end{array} \right\} \quad (1.1.2)$$

The formulation of a maximization problem is of course similar. Every value for $\mathbf{X}$ is called a *solution* of the problem. If a solution satisfies the constraints it is called a *feasible solution*. That solution which satisfies the constraints and yields the maximum or minimum value for the objective function is called the *optimal solution*. We will mostly mark this solution with an asterisk, so $\mathbf{X}^*$ is the optimal solution.

In this book the transport network optimization problem is dealt with. This means that the decision variables are variables defined on a transport network and that some of the constraints serve only to describe the relationships of a transport network. The concept 'transport network' will be described in the second part of this chapter and in the second chapter.

The definition of the objective function, the decision variables and the constraints forms the most essential part of an optimization problem. They will be discussed in the third chapter. The techniques for finding the optimal solution are discussed in the fourth and fifth chapters. In the subsection now following we will give a very brief introduction to some parts of the theory of optimization, as far as it is relevant to the problems discussed in this book. There are very many excellent textbooks and papers on optimization; it forms one of the major parts of pure and applied mathematics. For a general survey see, for instance, the books of Hadley (1962 and 1964) and Timman (1966).

1.1.2 Optimization Techniques

Since the middle of the eighteenth century the differential calculus has been known and applied to solve optimization problems. The simplest problem is that of optimizing a function of one variable without constraints:

$$\min_{x} F(x)$$

It can be proved easily that for differentiable functions it is a necessary condition for the minimum (x^*) that the first derivative of F with respect to x equals zero in that point:

$$\left(\frac{\mathrm{d}F}{\mathrm{d}x}\right)_{x=x^*} = 0$$

This condition is not sufficient; it is also necessary that the second derivative is non-negative in that point:

$$\left(\frac{\mathrm{d}^2F}{\mathrm{d}x^2}\right)_{x=x^*} \geqslant 0$$

If the second derivative is greater than zero the solution is the minimum. If this second derivative equals zero the solution may be a minimum, but in that case further inspection is needed. The same as above can be said for a maximization problem. Only the second derivative needs to be non-positive.

A function may have more maxima and/or minima. A minimum is then defined as a solution for which the value of the objective function is just lower than the values for the solutions in a neighbourhood of the minimum solution. Such a minimum is called a *local minimum*. The absolutely lowest value of the objective function is reached by the *global minimum*. So the global minimum is the minimum of all local minima. The same can be said for maxima.

Besides the classical approach of the differential calculus there exists the possibility of using a search technique to find the optimum in the case of one decision variable. These techniques are mentioned in Section 10.1.

There are two special and important kinds of functions: convex and concave functions. A convex function has the following property:

$$F\{\lambda x_1 + (1 - \lambda)x_2\} \leqslant \lambda F(x_1) + (1 - \lambda)F(x_2) \qquad \text{with } 0 \leqslant \lambda \leqslant 1$$

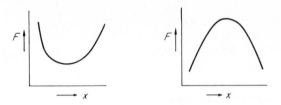

(a) A strictly convex function   (b) A strictly concave function

Figure 1.1.1   A strictly convex and a strictly concave function

A function is called strictly convex if

$$F\{\lambda x_1 + (1 - \lambda)x_2\} < \lambda F(x_1) + (1 - \lambda)F(x_2) \qquad \text{with } 0 < \lambda < 1$$

A function is called concave if

$$F\{\lambda x_1 + (1 - \lambda)x_2\} \geqslant \lambda F(x_1) + (1 - \lambda)F(x_2) \qquad \text{with } 0 \leqslant \lambda \leqslant 1$$

and strictly concave if

$$F\{\lambda x_1 + (1 - \lambda)x_2\} > \lambda F(x_1) + (1 - \lambda)F(x_2) \qquad \text{with } 0 < \lambda < 1$$

Of course the negative of a convex function is concave and vice versa. Moreover it can be seen very easily that the sum of a number of (strictly) convex functions is again a (strictly) convex function and that the same is true for (strictly) concave functions. Figure 1.1.1 shows a (strictly) convex and a (strictly) concave function. A linear function ($F = ax + b$) is, of course, both convex and concave.

It can be seen that, if a convex function is twice differentiable, the second derivative is never negative; while the second derivative of a differentiable concave function is never positive. It can be proved that if a convex function has a local minimum it is also a global one and that if a concave function has a local maximum it is a global one. For strictly convex or concave functions there will be only one unique optimal solution. If there are more optimal solutions for not-strictly convex or concave functions there is an infinite number of optimal solutions, all with the same value for the objective function. Moreover if the decision variable is only defined for a closed interval: $x_a \leqslant x \leqslant x_b$ then there is always a minimum for a convex function and a maximum for a concave function, and, furthermore, there is also a maximum for the convex function and a minimum for the concave function, which will be situated on the border(s)

$$(x^* = x_a \text{ or/and } x^* = x_b).$$

For functions of more variables the differential calculus can be used again to yield the conditions for an optimal solution. Here the partial derivatives play a role. A partial derivative of a function $F(x_1, \ldots, x_n)$ with respect to x_i

is defined as:

$$\frac{\partial F}{\partial x_i} = \lim_{\Delta x \to 0} \frac{F(x_1, \ldots, x_i + \Delta x, \ldots, x_n) - F(x_1, \ldots, x_i, \ldots, x_n)}{\Delta x}$$

In the optimal solution all partial derivatives are equal to zero. This is only a necessary condition. Sufficient conditions for a maximum or a minimum can be given sometimes by (combinations of) the second derivatives; sometimes an inspection of the neighbourhood of the solution is needed. However, in the case of convex or concave functions (the convexity or the concavity of a function of more variables is defined in a similar way as for a function of one variable) the same conditions can be derived as for functions of one variable. So a stationary point (a point for which all partial derivatives equal zero) is always an optimum solution and a local optimum is a global one. Finally if the function is defined on a closed region there will be always an optimum and this optimum is unique when the function is strictly convex or strictly concave. All these things can be proved easily, but we refer to the standard literature for that.

To state the necessary and/or sufficient conditions for the optimum solution and to find that optimum solution are two different things. It can be a very complicated if not impossible task even to solve the equations for the stationary points. So special techniques have been developed to find the optimum. A very well-known method is the gradient method. The gradient of a function is a vector with the partial derivatives as elements:

$$\nabla F = \left\{ \frac{\partial F}{\partial x_1}, \ldots, \frac{\partial F}{\partial x_n} \right\}$$

The gradient is always directed towards the optimum (or in the very opposite direction), so $\mathbf{X} + \alpha \nabla F$ is closer to the optimum than $\mathbf{X}$ (with α chosen in the correct way positive or negative). So one can try to reach the optimum in a number of steps taken in the direction of the gradient.

Now we turn to optimization problems including constraints. The classical approach for this problem is the use of Lagrange multipliers. That is, the constrained problem:

$$\min_{\mathbf{X}} f(\mathbf{X}) \qquad \text{subject to } g(\mathbf{X}) = 0 \tag{1.1.3}$$

is converted into the unconstrained problem:

$$\min_{\mathbf{X}, \lambda} F(\mathbf{X}, \lambda) = f(\mathbf{X}) + \lambda g(\mathbf{X}) \tag{1.1.4}$$

It can be proved that the optimal solution of problem (1.1.4) is also the optimal solution of (1.1.3).

Using the concept of Lagrange multipliers it can be shown that some different problems have the same or very similar mathematical formulations. In this book we deal with problems about transport and investments in infrastructure. We will assume for a moment now that the total costs of transport consist of travel time costs, evaluated in monetary terms, and capital invested in roads. We will formulate several optimization problems in transport.

A few examples of these are given here for the case of fixed traffic flows on the different roads. The dimensions of the roads are the decision variables. We use the following variables:

$\mathbf{C}$ — dimensions (capacities) of the roads

$\mathbf{X}$ — traffic flows

$Z(\mathbf{X}, \mathbf{C})$ — travel time as a function of the dimensions and the traffic flows

$I(\mathbf{C})$ — capital invested as a function of the dimensions

k — factor, which makes it possible to compare (add) travel time costs and capital invested

We can formulate the following problems now:

(a) minimize the total costs:

$$\min_{\mathbf{C}} F = kZ(\mathbf{X}, \mathbf{C}) + I(\mathbf{C}) \tag{1.1.5}$$

(b) minimize the travel time with a fixed budget for investments I^0:

$$\min_{\mathbf{C}} Z(\mathbf{X}, \mathbf{C}) \text{ subject to } I(\mathbf{C}) = I^0 \tag{1.1.6}$$

(c) minimize the capital invested for a given travel time (level of service):

$$\min_{\mathbf{C}} I(\mathbf{C}) \qquad \text{subject to } Z(\mathbf{X}, \mathbf{C}) = Z^0 \tag{1.1.7}$$

Using the concept of Lagrange multipliers we see that the problems (1.1.6) and (1.1.7) are very similar to the problem (1.1.5). The only difference is that the factor k is fixed in advance in problem (1.1.5) and is a Lagrange multiplier in the other two problems where it is then a result of the optimization. A little working-out will show this.

Writing problem (1.1.6) as an unconstrained minimization problem with the use of a Lagrange multiplier we get:

$$\min_{\mathbf{C}, \lambda} F(\mathbf{C}, \lambda) = Z(\mathbf{X}, \mathbf{C}) + \lambda(I(\mathbf{C}) - I^0) \tag{1.1.8}$$

If λ were not a decision variable it would be possible to divide relationship (1.1.8) by λ and to omit the last term I^0, which would be a constant then. Writing k for $1/\lambda$ relationship (1.1.8) becomes:

$$\min_{\mathbf{C}, k} F(\mathbf{C}, k) = kZ(\mathbf{X}, \mathbf{C}) + I(\mathbf{C})$$

The constrained minimization problem (1.1.7) can be written as the following unconstrained minimization problem:

$$\min_{C, \lambda} F(C, \lambda) = I(C) + \lambda(Z(X, C) - Z^0)$$

which can be rewritten as:

$$\min_{C, k} F(C, k) = kZ(X, C) + I(C) - kZ^0 \qquad (1.1.9)$$

The last term of (1.1.9) would be constant if k were not a decision variable; in that case it could be omitted. The similarity of the three problems (1.1.5), (1.1.6), and (1.1.7) is obvious now.

It can be proved that in the optimal solution the derivative of the objective function with respect to (the constant term in) the constraint equals the Lagrange multiplier. This has led to the interpretation of the Lagrange multiplier as a shadow-price of the constraint. So, for problem (1.1.6), λ can be interpreted as the price for capital and, for problem (1.1.7), λ forms the price for travel time. The interpretation thus resulting of the factor k as the travel time evaluation divided by the price of capital can also be derived directly from the statement of the problem (1.1.5).

Optimization problems involving more constraints can also be written as unconstrained optimization problems using Lagrange multipliers. In that case as many Lagrange mutlipliers are necessary as there are constraints. Finally, Lagrange multipliers can also be used in the case of inequality constraints.

For the cases just mentioned, with more constraints, the use of Lagrange multipliers is, generally speaking, only of theoretical interest. The systems of equations to be solved are mostly too complicated to produce a correct solution in a reasonable computation time. So a whole class of new optimization techniques had to be devised. These form the main developments of the theory of optimization and an important object of the whole field of operations research in the last decades. Problems and solution methods of this kind are sometimes also called programming problems and methods. The problems are usually run on an electronic computer.

The best-known programming problem is the linear programming problem in which the objective function and the constraints are all linear. The problem has the following general form:

$$\min_{x_j} \sum_{j=1}^{n} b_j x_j$$

subject to

$$\sum_{j=1}^{n} a_{ij} x_j <, =, > a_i^0 \qquad i = 1, \ldots, m$$

Many problems have been and will be formulated in this way. Also many network optimization problems can be stated and solved as a linear program, to such an extent that many authors refer to network theory as a special branch of linear programming (for instance Ford and Fulkerson, 1962).

The solution of a linear program has become a standard problem and almost all computer program libraries possess a standard software package for the linear programming problem. To solve a linear program the Simplex algorithm is used, which yields an exact optimum in a finite number of steps (unless there are no feasible solutions or there is an infinite number of optimal solutions).

When the objective function and/or constraints are not all linear the solution of the optimization problem becomes much more complicated. Certain techniques exist for certain kinds of problems. Quadratic programming exists for the case of linear constraints and a quadratic objective function; geometric programming exists for the case of constraints and an objective function consisting of terms that are products of powers of the decision variables; dynamic programming exists for the case of an objective function that is the sum of functions of only some of the decision variables and constraints that also involve only some of the variables. Dynamic programming will be discussed in Section 6.1.2.

A very important class of problems is formed by those problems for which the variables are restricted to be integer. This kind of problem is called integer programming or mixed-integer programming if the integrity constraint does not hold for every variable. A special class of integer programming problems is formed by those problems for which the decision variables can take only a finite number of values. These problems can always be reduced to problems in which the decision variables can only take the values 0 or 1. (For every positive integer number can be written as the summation of integer powers of 2:

$$l = \sum_{i=0}^{n} y_i 2^i \quad \text{with} \quad y_i = 0 \text{ or } 1.)$$

Then we speak of combinatorial problems. Theoretically such a combinatorial problem can always be solved by trying out all possibilities (exhaustive enumeration). In practice faster techniques are used such as dynamic programming (see Section 6.1.2) and branch and bound (see Section 4.3).

Sometimes the problems are so complicated and involve so many decision variables and/or constraints that the requirement of finding an optimal solution is relaxed and replaced by the desire to find a feasible solution that is fairly good. In such a case we speak of heuristic techniques or heuristic programming. We will treat some heuristic methods in Section 4.4.

Finally, one can try to reduce the size of the problem (the number of variables and/or constraints). Then we speak of aggregation or decomposition. This subject will be treated briefly in Section 4.5.

We will use some of the optimization techniques mentioned here on problems concerning transport in networks. But first we will continue by defining a network in the next section.

1.2 NETWORK FLOWS

The theory of graphs and network flows is an important branch of pure and applied mathematics. There exist many excellent books and papers on this subject, of which we will mention those by Busacker and Saaty (1965), Ford and Fulkerson (1962), Hu (1969) and Elmaghraby (1970).

1.2.1 Graphs and Networks

A *graph* consists of a set N of elements $i, j, \ldots$ together with a set L of ordered pairs ij of elements of N. The elements of N are called the *nodes*, vertices or (junction) points of the graph. The elements of L that connect the nodes, are called the *links*, arcs, branches, edges or lines of the graph. We will use the node-link terminology throughout this book. In Figure 1.2.1

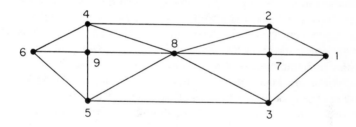

Figure 1.2.1   A graph

a graph is shown. The set N consists of the nodes $1, 2, 3, 4, \ldots, 9$ and the set L of the links $12, 21, 17, 71, 13, 31, 24$ and so on.

A graph can be connected or unconnected. In a connected graph every node can be reached along links of the graph by every other node. A finite graph consists of a finite number of nodes and links.

A graph is called linear when the links have no points in common other than the nodes. A graph is called planar if all the links connecting the nodes can be mapped into the plane such that the links have no points in common other than the nodes. We will use throughout graphs, that are connected, finite and linear and will, therefore, omit the epithets 'connected', 'linear' and 'finite' when speaking about graphs.

We define a link as an ordered pair of nodes. So a link has a certain orientation or direction. Sometimes the term directed link is used for an ordered pair of nodes as opposed to the term undirected link for an unordered pair of nodes. Unless it is explicitly stated otherwise we will be using ordered pairs of nodes throughout, so link ij is different from link ji.

A sequence of links $i_1i_2, i_2i_3, i_3i_4, \ldots, i_{n-1}i_n$ forms a *path*, a route or a chain. In Figure 1.2.1 for instance 1246 forms a path from 1 to 6. Usually the links in a path can be directed in either direction while the links in a chain are directed in the same direction. In this book however we assume a path consisting of links directed in the same direction. So i_1i_2, i_3i_2, i_3i_4 does not form a path according to our definition. If the first and the last node of a path coincide $(i_1i_2i_3i_4 \ldots i_ni_1)$ we speak about a *loop* or a cycle. In Figure 1.2.1 a loop is formed by the path 171.

A *tree* is a graph with no loops in which every node is connected to every other node. It can be proved by induction that a tree with n nodes consists of $(n-1)$ links. A subgraph of a graph is a graph consisting of a subset of the nodes of the original graph and a subset of the links of the original graph. A spanning tree is a subgraph such that every node of the original graph is in the tree.

We define a *network* as a graph of which the links and/or the nodes have some quantitative characteristics. Some authors do not make the distinction between graphs and networks. Following, for instance, Elmaghraby, we do. We use the concept 'graph' for the definition of the purely structural relationships between the nodes; while in our definition a 'network' bears in addition quantitative characteristics of the links and/or the nodes.

A very popular quantitative characteristic of a link is the length of a link. We will use the symbol d_{ij} for the length of the link from i to j. The length of a path is the summation of the lengths of all links contained in the path.

For the length of a path p from node a to node b we use the symbol d^{pab}. So we get:

$$d^{pab} = \sum_{\substack{ij \in p, \, p \text{ is a path from } a \text{ to } b \\ ij \in L}} d_{ij} \qquad (1.2.1)$$

where the notation $ij \in L$ means link ij belongs to the set L of all links.

Of course it is possible to look for the path p among the set Pa^{ab} of all possible paths from a to b with the shortest value for the length. We use the symbol d^{*ab} for the length of this shortest path:

$$d^{*ab} = \min_{p \in Pa^{ab}} d^{pab} \qquad (1.2.2)$$

where the notation means: look for the minimum value of d^{pab} by choice of the path p, while p belongs to the set Pa^{ab}.

It is also possible, and sometimes profitable, to define the length of the shortest path in a recursive way:

$$d^{*ab} = \min_{\substack{i \\ (ib \in L)}} (d^{*ai} + d_{ib}) \qquad (1.2.3)$$

In the seventh chapter of this book we will give an ample treatment of the problem of finding the shortest path in a network.

Besides the shortest path problem there exist in network theory many other well-known problems which can be defined now: the problem of the minimum spanning tree, the problem of the longest path, better known by such names as 'Critical Path Method' (CPM) or 'Program Evaluation and Review Technique' (PERT), the travelling-salesman problem and so on.

The minimum spanning tree of a network is that spanning tree with the lowest value for the sum of the lengths of the links contained in the tree. The longest or the critical path is that path among all possible paths connecting the nodes with the maximum value for the length. In the models like CPM or PERT the links or nodes represent some activities that are necessary to complete some project. Finally the travelling-salesman problem is the problem of the salesman who starts from one city, visits each of all the other cities only once and returns to his starting position having travelled the minimum distance.

Although a study of the problems mentioned in the paragraph above could also be very useful for the class of problems of network optimization discussed in this book, we will not deal with them here.

Two other important quantitative characteristics of a link will be introduced now: the dimension or capacity c_{ij} of a link and the flow x_{ij} on a link. Both are assumed to be non-negative real numbers:

$$c_{ij} \geqslant 0 \quad \text{for all } ij \in L \tag{1.2.4}$$

$$x_{ij} \geqslant 0 \quad \text{for all } ij \in L \tag{1.2.5}$$

In most network studies another restriction is that the flow cannot exceed the capacity of a link ($x_{ij} \leqslant c_{ij}$). In this book we will not use this restriction. Instead we will assume the length of a link to be a function of the dimension and the flow of the link:

$$d_{ij} = d_{ij}(x_{ij}, c_{ij}) \quad \text{for all } ij \in L \tag{1.2.6}$$

This function generally has the property of being a monotonically decreasing function of c_{ij} and a monotonically increasing function of x_{ij}. Moreover the value of d_{ij} tends to become infinity if x_{ij} exceeds c_{ij} by too much. So we have the following properties:

$$\left. \begin{array}{c} \dfrac{\partial d_{ij}}{\partial c_{ij}} \leqslant 0 \\[2ex] \dfrac{\partial d_{ij}}{\partial x_{ij}} \geqslant 0 \\[2ex] \lim_{\frac{x_{ij}}{c_{ij}} \to \infty} d_{ij} = \infty \end{array} \right\} \quad \text{for all } ij \in L \tag{1.2.7}$$

We will further discuss the concept of network flow in the following sub-section.

1.2.2 Flows in Networks

In the preceding subsection we introduced the concept of a flow in a network. We will give a full definition of a flow below. To do that we must define two special nodes in the network: the node a, which we call the origin or the source, and the node b, which we call the destination or the sink. All other nodes are called intermediate nodes. As the flow in the network we consider now all vehicles or persons travelling on the several links in the network from the origin a to the destination b. This formulation is made precise by defining a flow as a set of non-negative real numbers, satisfying the following constraints:

$$\sum_{\substack{i \\ (ij \in L)}} x_{ij} - \sum_{\substack{k \\ (jk \in L)}} x_{jk} \begin{cases} = 0 & \text{for all } j \neq a \text{ or } b\,;\, j \in N \\ = -x & \text{if } j = a \\ = x & \text{if } j = b \end{cases} \tag{1.2.8}$$

Relationship (1.2.8) says that the net flow is zero for every node except for the origin and the destination. So these relationships are called the *conservation laws*.

One of the problems most frequently dealt with in network theory is the problem of defining the maximum flow through a network. To get a meaning-ful statement of the problem it is necessary that the restrictions $x_{ij} \leqslant c_{ij}$ hold. Now the question is to find the maximum value of the total flow from a to b (x in relationship (1.2.8) that can flow through the network. As said before we are not using the restrictions $x_{ij} \leqslant c_{ij}$ in this book; we therefore will not deal with the maximum flow problem.

Besides the uniterminal flow there exists the multiterminal flow with many origins and many destinations. The conservations laws are very similar to those of relationship (1.2.8):

$$\sum_{\substack{i \\ (ij \in L)}} x_{ij} - \sum_{\substack{k \\ (jk \in L)}} x_{jk} \begin{cases} = 0 & \text{if } j \in N^I \\ = -x^j & \text{if } j \in N^O \\ = x^j & \text{if } j \in N^D \end{cases} \tag{1.2.9}$$

in which:

N^I —set of intermediate nodes
N^O—set of origins
N^D—set of destinations

To get a feasible solution it is of course necessary that:

$$\sum_{i \in N^O} x^i = \sum_{j \in N^D} x^j$$

It is possible to replace a multiterminal flow by a uniterminal one by connecting all origins to a super-origin and all destinations to a super-destination and assuming that the total flow originates in the super-origin and terminates in the super-destination. All original origins and destinations become intermediate nodes now.

We can also define the maximum-flow problem for the multiterminal flow. Moreover some more very well-known problems, such as the standard transportation, or Hitchcock, problem and the transhipment problem, can be defined now. For the transportation problem there exist a set of origins having a supply of goods x^i and a set of destinations having a demand of goods x^j and connections ij between the origins and destinations. To transport one unit from i to j, t_{ij} cost units are needed.

The problem is now to transport the goods from the origins to the destinations at minimum cost:

$$\min_{x_{ij}} \sum_{ij \in L} x_{ij} t_{ij}$$

subject to:

$$\sum_{\substack{k \\ (ik \in L)}} x_{ik} \leqslant x^i \quad \text{for all } i \in N^O$$

$$\sum_{\substack{l \\ (lj \in L)}} x_{lj} \geqslant x^j \quad \text{for all } j \in N^D \tag{1.2.10}$$

To be able to get a feasible solution it is of course necessary that:

$$\sum_{i \in N^O} x^i \geqslant \sum_{j \in N^D} x^j$$

Ford and Fulkerson showed that this problem has some relationship to a maximum flow problem.

Besides the standard transportation problem (1.2.10) there exists the transhipment problem. Here too the costs have to be minimized but as well as origins and destinations there are also intermediate nodes. So the restrictions of relationship (1.2.9) hold in that case.

There are many other well-known problems with the same or very similar mathematical formulations, for example the problem of assigning candidates to jobs and other allocation problems. Again a study of the methods of solving these problems can be very fruitful for our problem, but we will not deal with them here.

1.2.3 Multicommodity Flows

In the preceding subsection we met the network problem where there were many origins and many destinations. But the flow from any arbitrary origin could be sent to any arbitrary destination and conversely (provided that the nodes are connected). Also it was very easy to convert the multi-terminal problem into an uniterminal one.

If we make the restriction that the flow from certain origins must be sent to certain destinations, we get another class of problems. These are the so-called multicommodity flows. A special case of multicommodity flows is the case in which every flow goes from just one origin to just one destination. For the total flow from node a to node b we use the notation x^{ab}. The ordered pair ab, consisting of an origin node a and a destination node b, is called a (transport) *relation* and P forms the set of all these relations. Note that the relation ab is only sensible when a is connected to b by a path in the network.

The conservation laws of the network flows now take on a more complicated form:

$$\sum_{\substack{i \\ (ij\in L)}} x_{ij}^{ab} - \sum_{\substack{k \\ (jk\in L)}} x_{jk}^{ab} \begin{cases} = 0 & \text{for all } j \neq a \text{ or } b\,;\, j\in N \\ = -x^{ab} & \text{if } j = a \\ = x^{ab} & \text{if } j = b \end{cases} \tag{1.2.11}$$

$$\text{for all } ab \in P$$

in which x_{ij}^{ab} means the flow of relation (commodity) ab on link ij. Note that x_{ia}^{ab} does not exist when there are no loops.

In the case of multicommodity flows, too, we have the requirement that the flows can be never negative:

$$x_{ij}^{ab} \geqslant 0\,; \quad \begin{array}{l} \text{for all } ab \in P \\ \text{for all } ij \in L \end{array} \tag{1.2.12}$$

It is important to see how many relationships for how many variables the conservation laws involve. Therefore we define:

n_N—number of nodes
n_L—number of links
n_P—number of transport relations.

The total number of variables is then $(n_L + 1)n_P$ and the number of relationships $n_P n_N$. These can be enormous numbers. In the Dutch Integral Transportation Study where 351 origins and 351 destinations yielded approximately 125,000 relations, we had to cope with approximately 2,000 nodes and 6,000 links. Application of the conservation laws meant that 250 million relationships for 750 million variables had to be considered, to which another 750 million relationships had to be added because of non-negativity requirements. These enormous numbers of variables and relationships are the intrinsic difficulty of the subject dealt with in this book.

We assume that the flows of different commodities on a link can be added to each other. So we get the following relationship for every link:

$$x_{ij} = \sum_{ab \in P} x_{ij}^{ab}; \quad \text{for all } ij \in L \qquad (1.2.13)$$

Having this relationship (1.2.13) we can also use the relationship between the length and dimension and flow of a link as given in relationship (1.2.6) for multicommodity flows in networks.

It is possible to write the relationships (1.2.11), (1.2.12) and (1.2.13) in a shorter way using vectors and matrices. We have already defined a vector in Section (1.1.1) as a one-dimensional array of real numbers: $\mathbf{X} = (x_1, \ldots, x_n)$. A matrix is a rectangular array of real numbers. A matrix possesses rows and columns. We will generally indicate a matrix with a capital letter. It is obvious that we can see a vector as a matrix with only one row or column. To be able to simplify the writing-down of the relationships we must define the product of a matrix with a vector. The product of a matrix and a vector (whlch must have as many elements as there are columns in the matrix) is a vector with as many elements as there are rows in the matrix, for which the mth element is the sum of the products of the elements of the mth row and the elements of the vector. The element of the nth column is always multiplied by the nth element of the vector.

We can now write down the relationships (1.2.11) and (1.2.13) in the form:

$$A\mathbf{X} = \mathbf{0} \qquad (1.2.14)$$

and the relationship (1.2.12) in the form:

$$\mathbf{X} \geqslant \mathbf{0} \qquad (1.2.15)$$

The composition of the matrix A and the vector $\mathbf{X}$ is shown in Figure 1.2.2.

We will quite often use the relationships (1.2.14) and (1.2.15), referring to them as the 'network constraints'. The same mathematical expression can be used to formulate another set of constraints. For the elements x^{ab} of the vector $\mathbf{X}$ have a special meaning. They give the flow for every relation. When working with transport networks we express the flow in trips, i.e. the number of persons or vehicles travelling from an origin to a destination. Thus we use the name '*trip-matrix*' for the elements of x^{ab}. To use the word matrix for a part of a vector is perhaps a little confusing, but it only refers to the way the elements are written down. The columns of the trip-matrix are written down below each other and in that way form a part of the vector $\mathbf{X}$. There are now two important ways of working with the trip-matrix, which will be treated further in the second chapter. The trip-matrix can be fixed in advance or it can be a function of the different variables of the network. In the first case we generally say that the relationships (1.2.14) and (1.2.15) constitute both the network constraints and the fixed trip-matrix. In the

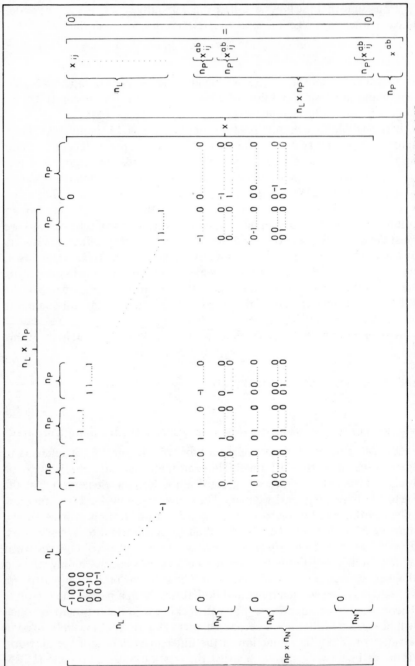

Figure 1.2.2   The conservation laws and additions in matrix–vector notation [relationship (1.2.14)]

econd case the relationships are said to constitute only the network con-
traints.

It is possible, and sometimes advantageous, to put the conservation laws
and other constraints in another way, using the concept of paths. We intro-
luce the variable:

x_{ij}^{pab}—flow on link ij of the relation ab using path p (from a to b).

he conservation laws now become:

$$x_{ij}^{pab} = x_{kl}^{pab} \begin{cases} \text{for all } ij \in p \text{ and } kl \in p \\ \text{for all } p \in Pa^{ab} \\ \text{for all } ab \in P \end{cases} \tag{1.2.16}$$

nd:

$$\sum_{p \in Pa^{ab}} x_{aj}^{pab} = x^{ab}; \quad \text{for all } ab \in P \tag{1.2.17}$$

Of course the non-negativity restriction also holds:

$$x_{ij}^{pab} \geqslant 0 \tag{1.2.18}$$

nd also:

$$x_{ij} = \sum_{\substack{p \in Pa^{ab} \\ ab \in P}} x_{ij}^{pab} \tag{1.2.19}$$

t is also possible to write down the relationships (1.2.16) up to (1.2.19) in
ne form $AX = 0, X \geqslant 0$; only the matrix A and the vector X have many
nore elements than when the concept of paths is not used.

It is possible to use a set of weaker constraints for the conservation laws
•r multicommodity flows. In that case no distinction is made between the
different flows from one origin to the different destinations or conversely
om the different origins to the same destination. This seems reasonable
•r the description of traffic flows. Using N^O again for the set of all origins
nd N^D for the set of all destinations, the conservation laws become:

$$\sum_{\substack{i \\ (ij \in L)}} x_{ij}^a - \sum_{\substack{k \\ (jk \in L)}} x_{jk}^a \begin{cases} = 0 & \text{for all } j \neq a; j \in N \text{ but } j \notin N^D \\ = -x^a & \text{if } j = a \\ = x^{ab} & \text{if } j = b \text{ for all } b \in N^D \end{cases}$$

$$\text{for all } a \in N^O \tag{1.2.20}$$

where x_{ij}^a is the flow on link ij originating in a, or:

$$\sum_{\substack{i \\ (ij \in L)}} x_{ij}^b - \sum_{\substack{k \\ (jk \in L)}} x_{jk}^b \begin{cases} = 0 & \text{for all } j \neq b; j \in N \text{ but } j \notin N^O \\ = -x^{ab} & \text{if } j = a \text{ for all } a \in N^O \\ = x^b & \text{if } j = b \end{cases}$$

for all $b \in N^D$ (1.2.21)

where x_{ij}^b is the flow on link ij destinating in b. To be complete the following constraints are also necessary:

$$x_{ij}^a \geqslant 0 \qquad \begin{matrix} \text{for all } a \in N^O \\ \text{for all } ij \in L \end{matrix} \qquad (1.2.22)$$

$$x_{ij} = \sum_{a \in N^O} x_{ij}^a \quad \text{for all } ij \in L \qquad (1.2.23)$$

or:

$$x_{ij}^b \geqslant 0 \qquad \begin{matrix} \text{for all } b \in N^D \\ \text{for all } ij \in L \end{matrix} \qquad (1.2.24)$$

$$x_{ij} = \sum_{b \in N^D} x_{ij}^b \quad \text{for all } ij \in L \qquad (1.2.25)$$

Even though when this formulation is used the number of variables and constraints is much less than when working with the flow x_{ij}^{ab}, the number is still very large. It is now also possible, of course, to write down the relationships in the form $A\mathbf{X} = \mathbf{0}, \mathbf{X} \geqslant \mathbf{0}$.

The relationships (1.2.11) up to (1.2.13), or (1.2.14) and (1.2.15) or (1.2.16) up to (1.2.19), or (1.2.20), (1.2.21), and (1.2.23), or (1.2.21), (1.2.24) and (1.2.25) together with the relationship (1.2.6) for the length of the links, are a necessary part of the general description of a transport network. In a transport network the flows consist of traffic flows of persons or vehicles. The length of the links can have several implications, e.g. attractiveness or unattractiveness to travellers, or higher or lower costs to society.

The relationships mentioned form the technical relationships of a transport network and are sometimes called the supply characteristics of transport. The other aspect of a transport network is the behaviour of the travellers in the network, how they respond to the several variables of the network. This constitutes the demand for transport. These behavioural relationships will be discussed in the second chapter. It has become clear that we have now defined an important class of relationships of relevance to the problem treated in this book. Usually these relationships will constitute (a part of) the constraints of the optimization problem.

REFERENCES

Busacker, R. G., and Saaty, T. L. (1965). *Finite Graphs and Networks*, McGraw-Hill, New York.

Elmaghraby, S. E. (1970). The theory of networks and management science. *Management Science*, 17, Nos. 1 and 2 (September and October).

Ford, L. R., and Fulkerson, D. R. (1962). *Flows in Networks*, Princeton University Press, Princeton N.J.

Hadley, G. (1962). *Linear Programming*, Addison-Wesley, Reading, Mass.

Hadley, G. (1964). *Nonlinear and Dynamic Programming*, Addison-Wesley, Reading Mass.

Hu, T. C. (1969). *Integer Programming and Network Flows*, Addison-Wesley, Reading Mass.

Timman, R. (1966). *Optimaliseren van Funkties en Funktionalen*, Technische Hogeschool Delft, Onderafdeling der Wiskunde, Delft.

2

Transportation

2.1 TRANSPORTATION—DESCRIPTION OF THE SYSTEM

2.1.1 A Transport Network

We define transportation as the transfer of persons and/or goods, in a vehicle or otherwise, between geographically separate places. The collective movement of vehicles or persons between those places or within a certain area is called traffic.

For the movement of the vehicles or persons, the transport infrastructure is used. The transport infrastructure can be divided into three components.

(a) the fixed objects as highways, secondary roads, city arterials, streets, railways, airports, canals, pipelines and so on, all with certain characteristics;
(b) the vehicles using the fixed infrastructure;
(c) the organizational system necessary to ensure that the vehicles and the fixed infrastructure are well used.

In the mathematical models we will use, the transport infrastructure system is abstracted into a *transport network* consisting of nodes and links with certain characteristics. The relationships for a network given in Section 1.2 hold for such a network: the relationships for the lengths of the links (relationship (1.2.6)) and the so-called network constraints (for multicommodity flows (relationships (1.2.11), (1.2.12) and (1.2.13)), which say that the network flows are never negative, additive to each other and conservative in every node except the origins and destinations. Moreover the nodes and links may possess further characteristics, for instance a link may only be used by a certain mode of transport.

We may distinguish the system 'transportation' into the supply of transport and the demand for transport. The transport infrastructure forms the supply of transport and the demand for transport is formed by the persons and goods willing to be transferred together with the underlying causes and behavioural relationships. We will describe the basic behavioural relationships for the demand for transport in the next subsection.

2.1.2 The Demand for Transport

In building a model, a hypothesis or set of hypotheses forming a 'theory' is formulated and it is then tested with quantitative data of the phenomena to be modelled. If the hypothesis or set of hypotheses is plausible and consistent and cannot be rejected by the data, it may be said that the phenomena have been described and explained. We will now give a brief description of the basic relationships for the model of the demand for transport. There is an extensive literature on transport network models. General surveys are given, for instance, by Overgaard (1966), Wohl and Martin (1969) and Hamerslag (1970 and 1972).

For reasons of simplicity we focus on person-trips in which the traveller himself is assumed to take the decisions about the trip. The relationships describing the demand for good-transport however can be developed in a similar way. At some points later in the book we will sometimes restrict ourselves to person-transport only while in other places both person- and good-transport are considered. But this alternation should not give rise to any difficulties.

To describe the demand for transport we follow the work of Hamerslag (1972) who bases his demand model on the general theory of micro-economics (see, for example Henderson and Quandt, 1958). Let us focus our attention on a trip. Related to this trip are certain costs and benefits. We distinguish between the costs and benefits to the user (the tripmaker) and those to society. They are not necessarily the same. For the description and explanation of the behaviour of the tripmaker only the costs and benefits to the users are assumed to be of relevance. The benefits of a trip from a to b arise in the first place from the linking of two activities practised in two geographically separate places, a and b (living in a, say, and working in b). Moreover the place of the activity may contribute to its value (living in a may be nicer than living in b). On the other hand there are the costs of travelling, such as fuel costs and time consumption, to be considered. An important fact is that these benefits and costs are personal; everybody has benefits and costs peculiar to himself.

Now, it is assumed that everybody maximizes the difference between his benefits and his costs:

$$\max_{m,p,b} (u^{ab} - t^{mpab}) \tag{2.1.1}$$

with:

u^{ab} —benefits of travelling from a to b

t^{mpab} —costs of travelling from a to b by mode m by route p

A man in a chooses the destination of a trip (b) (if $b = a$ he decides not to travel), the transport mode (m) and the route (p) for travelling so that the difference between the benefits and the costs, as he appreciates them, is maximized.

Because of the individual nature of the costs and benefits it happens that under the same conditions and at the same moment some people prefer one destination, mode and route and other people prefer others. Also a change in the costs will cause some people to change their destination, mode and/or route (for them $(u^{ab} - t^{m'p'ab'}) > (u^{ab} - t^{mpab})$), while other people (for whom $(u^{ab} - t^{mpab})$ is still the maximum) keep to their chosen destination, mode and/or route. This fact, *viz.* the number of people travelling in a particular relation by a particular mode and route is dependent on the values for the different costs and benefits, is usually formulated as a demand function:

$$x^{mpab} = f(u^{\alpha\beta}, t^{\mu\pi\alpha\beta}; \text{ all } \mu, \pi, \beta); \quad \text{for all } a \in N^0 \qquad (2.1.2)$$

with:

$\mu \in$ set of all travel modes
$\pi \in Pa^{ab}$
$\beta \in N^D$
x^{mpab} —number of trips from a to b by mode m on route p (for a certain time period).

This relationship (2.1.2) forms the model for the demand of transport. In transportation planning the demand functions are for practical reasons often replaced by a set of models (see, for example, Overgaard, 1966, Wohl and Martin, 1969 and Hamerslag, 1970 and 1972).

We get:
transport production (also called generation and attraction): the determination to travel or not to travel and the determination of the moment of travelling;
transport distribution: the determination of the destination (determination of b);
modal split: the choice of the mode of transport (determination of m);
route choice; the choice of the route (determination of p).

It is clear from the concept of maximizing the difference between benefits and costs (relationships (2.1.1) and (2.1.2)) that these four models in fact form one model. This, too, has been argued quite often (see for instance Wilson and Wagon, 1968 and Hamerslag, 1972). In practice this almost always means that the four models are executed in the order given followed by several feedbacks, converging, it is hoped, to a stable solution. Some attempts have been made to execute the models together, for instance in the SELNEC-study (Wilson and Wagon, 1968) where the distribution and the modal split are executed simultaneously in a model based on the theory of Wilson (1967) of maximizing the probability of the state of the system.

Another important feature is the fact that the benefits and the costs are influenced by the decisions of other tripmakers. To illustrate what we mean let us suppose for a moment that everybody goes to b to work. Now on the benefits side the effect will be that there are not enough jobs at b to go round, while on the costs side the increased traffic flow will cause congestion, which in its turn will make the costs per unit on that particular link go up. The situation in which the result of one's decision depends on the decisions of other people has been studied in the field of game theory (see von Neumann and Morgenstern, 1947). This game theory has been applied for instance to the route choice model (Charnes and Cooper, 1966 and Colony, 1970; also the paradox of Braess mentioned later in this book can be understood better if the concept of games is used).

2.1.3 Equilibrium in a Transport Network

After the demand and the supply function have been obtained the search is for the intersection. This is, of course, one of the main problems of transportation planning and almost all books and papers about modelling in transportation deal with this question.

For the practical working out of this question we refer to the many excellent books and papers on this subject (see, for instance, Overgaard (1966), Wohl and Martin (1969) and Hamerslag (1972). In this section we will deal with some theoretical aspects, concerning the existence and the uniqueness of a solution. To this end we will use very simple functions. Moreover we will introduce an approach also used later on in this book (when dealing with the derivation of the algorithm of the stepwise assignment according to the least marginal objective function for the network optimization problem).

The theory outlined in the next paragraphs was expounded by Beckmann, McGuire and Winsten in 1956, and was then based on conditions for an optimal solution for optimization problems given by Kuhn and Tucker (1951). Murchland (1969) draws attention to the work of Beckmann, *et al.* and refers at the same time to the theoretical work of Rockafellar (1967) on optimization conditions. We will use here an approach suggested by Timman (1966).

In Section 2.1.2 we gave the demand function (2.1.2) in which the total number of trips between two places is a function of the users' benefits and costs of travelling between these two and all other places. We assume now a simpler demand function in which the total number of trips between two places is only a function of the (average users') costs of travelling between these two places:

$$x^{ab} = x^{ab}(t^{ab}): \quad \text{for all } ab \in P \tag{2.1.3}$$

In this relationship t^{ab} are the general costs to the user for the trip from a to b. These costs are formed by the summation of the costs on the links of the path from a to b with the shortest length (in terms of general costs to the user), as defined in relationship (1.2.2).

We also define the inverse of the demand function:

$$t^{ab} = g^{ab}(x^{ab}); \quad \text{for all } ab \in P \qquad (2.1.4)$$

Now we formulate the principle of individual cost minimization with respect to the route choice. The result of this minimization is that everybody travelling from a to b chooses the path from a to b, that has the lowest value for the users' costs. This was already stated in 1841 by Kohl. It is very clearly formulated by Wardrop (1952) in his so-called 'second principle': 'The journey times on all the routes actually used are equal, and less than those which would be experienced by a single vehicle on any unused route'. (Although Wardrop mentions this principle first in his publication, p. 345, it seems always to be referred to as his 'second principle'.)

In our formulation this becomes (reading users' costs for journey times):

$$\left.\begin{aligned} &\text{if } x_{ij}^{ab} > 0 \text{ then } t^{ai} + t_{ij} = t^{aj} \quad \text{for all } ab \in P \\ &\text{if } t^{ai} + t_{ij} > t^{aj} \text{ then } x_{ij}^{ab} = 0 \quad \text{for all } ij \in L \end{aligned}\right\} \qquad (2.1.5)$$

On the supply side of the transport model we have the technical relationships of the network consisting of the network constraints (1.2.16) up to (1.2.19) and relationships for the users' costs on the links as a function of the flow:

$$t_{ij} = t_{ij}(x_{ij}) \quad \text{for all } ij \in L \qquad (2.1.6)$$

We impose two requirements on the functions $t_{ij}(x_{ij})$ and $g^{ab}(x^{ab})$:

$t_{ij}(x_{ij})$ is a non-decreasing function of x_{ij} for all $ij \in L$
$g^{ab}(x^{ab})$ is a non-increasing function of x^{ab} for all $ab \in P$

These requirements are satisfied in all normal real situations.

We now formulate the equilibrium problem in a transport network:

'Find an **X** such that the conditions (2.1.4), (2.1.5), (2.1.6) and (1.2.16) up to (1.2.19) are satisfied.' (2.1.7)

So a vector of flows is sought which satisfies the functions for the demand for and supply of transport.

For proof that a solution to this problem exists and that this solution is unique, it is shown that the conditions (2.1.4) and (2.1.5) form the necessary and sufficient conditions for the optimal solution of a minimization problem with the conditions (2.1.6) and (1.2.16) up to (1.2.19) as constraints.

First we define the objective function for that minimization problem:

$$F = \sum_{ab \in P} \int_0^{x^{ab}} - g^{ab}(x)\, dx + \sum_{ij \in L} \int_0^{x_{ij}} t_{ij}(x)\, dx \qquad (2.1.8)$$

where

$$x^{ab} = \sum_{p \in Pa^{ab}} x_{aj}^{pab} \qquad (1.2.17)$$

$$x_{ij} = \sum_{\substack{p \in Pa^{ab} \\ ab \in P}} x_{ij}^{pab} \qquad (1.2.19)$$

So, in fact, only the variables x_{ij}^{pab} are used.

From the requirements imposed on the functions $t_{ij}(x_{ij})$ and $g^{ab}(x^{ab})$ we see that the objective function is a convex function of x_{ij}^{pab} while x_{ij}^{pab} is also defined on a closed region. So an optimal solution does exist for the minimization problem and this solution is unique for the case where the objective function is strictly convex or otherwise there may be an infinite number of optimal solutions with the same value for the objective function.

In order to make it quite clear we state the minimization problem fully below:

$$\min_{x_{ij}^{pab}} F = \sum_{ab \in P} \int_0^{x^{ab}} - g^{ab}(x)\, dx + \sum_{ij \in L} \int_0^{x_{ij}} t_{ij}(x)\, dx \qquad (2.1.9)$$

subject to:

$$x_{ij}^{pab} = x_{kl}^{pab} \quad \begin{cases} \text{for all } ij \in p \text{ and } kl \in p \\ \text{for all } p \in Pa^{ab} \\ \text{for all } ab \in P \end{cases} \qquad (1.2.16)$$

$$x_{ij}^{pab} \geqslant 0 \quad \begin{cases} \text{for all } ij \in L \\ \text{for all } p \in Pa^{ab} \\ \text{for all } ab \in P \end{cases} \qquad (1.2.18)$$

Suppose $\mathbf{X}^*$ is the optimal solution, then the following relationship holds:

$$F(\mathbf{X}^*) \leqslant F(\mathbf{X}^* + \Delta\mathbf{X}) \qquad (2.1.10)$$

in which:

$$\mathbf{X}^* + \Delta\mathbf{X} \text{ (with elements } x^*{}_{ij}^{pab} + \Delta x_{ij}^{pab})$$

is a feasible solution in the 'neighbourhood' of $\mathbf{X}^*$.

Because $\mathbf{X}^* + \Delta\mathbf{X}$ is a feasible solution the following relationships hold for Δx_{ij}^{pab}:

from relationship (1.2.16):

$$\Delta x_{ij}^{pab} = \Delta x_{kl}^{pab} \quad \begin{cases} \text{for all } ij \in p \text{ and } kl \in p \\ \text{for all } p \in Pa^{ab} \\ \text{for all } ab \in P \end{cases} \quad (2.1.11)$$

from relationship (1.2.18):

$$\Delta x_{ij}^{pab} \geqslant 0 \quad \text{if } x*_{ij}^{pab} = 0 \quad \begin{cases} \text{for all } ij \in L \\ \text{for all } p \in Pa^{ab} \\ \text{for all } ab \in P \end{cases} \quad (2.1.12)$$

Expanding $F(\mathbf{X}^* + \Delta\mathbf{X})$ into a Taylor series we get:

$$F(\mathbf{X}^* + \Delta\mathbf{X}) = F(\mathbf{X}^*) + \sum_{\substack{p \in Pa^{ab} \\ ab \in P \\ ij \in L}} \left(\frac{\partial F}{\partial x_{ij}^{pab}} \right)_{\mathbf{X} = \mathbf{X}^*} \Delta x_{ij}^{pab}$$

$$+ \text{ higher order terms} \quad (2.1.13)$$

Because F is a convex function of $\mathbf{X}$ the sum of the higher order terms is always positive. We further suppose that $\Delta\mathbf{X}$ is so small that the sum of the higher order terms can be neglected (but is still positive); so we can write the inequality (2.1.10) as

$$\sum_{\substack{p \in Pa^{ab} \\ ab \in P \\ ij \in L}} \left(\frac{\partial F}{\partial x_{ij}^{pab}} \right)_{\mathbf{X} = \mathbf{X}^*} \Delta x_{ij}^{pab} \geqslant 0 \quad (2.1.14)$$

Before continuing we look at the derivatives $\partial F / \partial x_{ij}^{pab}$. If $i \neq a$ the first term of the objective function is constant with respect to x_{ij}^{pab} so the derivative becomes:

$$\frac{\partial F}{\partial x_{ij}^{pab}} = \frac{d}{dx_{ij}} \left(\int_0^{x_{ij}} t_{ij}(x) \, dx \right) \frac{\partial x_{ij}}{\partial x_{ij}^{pab}} \quad \text{for } i \neq a$$

Using equation (1.2.19) and differentiating we get:

$$\frac{\partial F}{\partial x_{ij}^{pab}} = t_{ij}(x_{ij}) \quad \text{for } i \neq a \quad (2.1.15)$$

If $i = a$ we get:

$$\frac{\partial F}{\partial x_{aj}^{pab}} = \frac{d}{dx^{ab}} \left(\int_0^{x^{ab}} - g^{ab}(x) \, dx \right) \frac{\partial x^{ab}}{\partial x_{aj}^{pab}} + \frac{d}{dx_{aj}} \left(\int_0^{x_{aj}} t_{ij}(x) \, dx \right) \frac{\partial x_{aj}}{\partial x_{aj}^{pab}}$$

Using equations (1.2.17) and (1.2.19) and differentiating we get:

$$\frac{\partial F}{\partial x_{aj}^{pab}} = -g^{ab}(x^{ab}) + t_{aj}(x_{aj}) \tag{2.1.16}$$

We now substitute the relationships (2.1.15) and (2.1.16) into relationship (2.1.14) and rearrange the terms in such a way that we sum over all paths:

$$\sum_{\substack{p \in Pa^{ab} \\ ab \in P}} \left(-g^{ab}\Delta x_{aj}^{pab} + \sum_{ij \in p} t_{ij}\Delta x_{ij}^{pab} \right) \geq 0 \tag{2.1.17}$$

Substituting equation (2.1.11) into relationship (2.1.17) we get:

$$\sum_{\substack{p \in P^{ab} \\ ab \in P}} \left(-g^{ab} + \sum_{ij \in p} t_{ij} \right) \Delta x_{aj}^{pab} \geq 0 \tag{2.1.18}$$

Combining relationships (2.1.12) and (2.1.18) we get the conditions:

$$\left.\begin{array}{l} \text{if } x^{*pab}_{aj} = 0 \quad \text{then } -g^{ab} + \sum_{ij \in p} t_{ij} \geq 0 \\[2mm] \text{if } x^{*pab}_{aj} > 0 \quad \text{then } -g^{ab} + \sum_{ij \in p} t_{ij} = 0 \end{array}\right\} \begin{array}{l} \text{for all } p \in Pa^{ab} \\[2mm] \text{for all } ab \in P \end{array} \tag{2.1.19}$$

But g^{ab} is independent of the path p so we get:

$$\left.\begin{array}{l} \sum_{ij \in p} t_{ij} = g^{ab} \quad \text{for all used paths from } a \text{ to } b \\[4mm] \sum_{ij \in p} t_{ij} \geq g^{ab} \quad \text{for all not used paths from } a \text{ to } b \end{array}\right\} \text{for all } ab \in P \tag{2.1.20}$$

From this we see that g^{ab} equals the length of the shortest path and so we can write the necessary and sufficient conditions for the optimal solution of the convex minimization problem as:

$$t^{ab} = g^{ab}(x^{*ab}) \quad \text{for all } ab \in P$$

and:

$$\begin{array}{ll} \text{if } x^{*ab} > 0 \quad \text{then } t^{ai} + t_{ij} = t^{aj} \text{ \Big\}} & \text{for all } ab \in P \\ \text{if } t^{ai} + t_{ij} > t^{aj} \quad \text{then } x^{*ab}_{ij} = 0 \text{ \Big\}} & \text{for all } ij \in L \end{array}$$

which equal the conditions (2.1.4) and (2.1.5) for the problem of equilibrium in a transport network.

In this section we have shown (assuming simple relationships) that there exists a solution for the state of equilibrium in a transport network and that this solution is a unique one (at any rate when the functions are strictly convex and strictly concave). Moreover we have shown that this state of equilibrium can be described as being the optimal solution of a minimization problem. Of course, the finding of this optimal solution is another thing. We will say something about it in this book but refer mainly to the books and papers about modelling in transportation research.

2.1.4 Traffic Assignment

2.1.4.1 The Traffic Assignment Problem

The traffic assignment problem is the problem of assigning the flows to the network so that the network constraints are satisfied. So the question is simply to find a feasible solution to the network constraints. The trip-matrix is of course fixed in this case.

It is obvious that there are many solutions to this general assignment or, to put it another way, that it is possible to apply further criteria for the assignment. One possibility is the assignment of the trip-matrix to the routes with minimum value for the users' costs, thus complying with the second principle of Wardrop. In this case the assignment problem equals the equilibrium problem discussed in the preceding section, only with a fixed demand:

$$x^{ab} = x^{0ab} \quad \text{for all } ab \in P \tag{2.1.21}$$

and no inverse of the demand function. Of course it is also now possible to state the equilibrium problem as an optimization problem.

2.1.4.2 Solution Methods for the Assignment Problem

An equilibrium assignment problem can be stated as an optimization assignment problem and vice versa. So we have only to solve one of these two problems. Bergendahl (1969) for instance uses an optimization technique (linear programming in this case) to apply an equilibrium assignment to a network. On the other hand, in Chapter 5 of this book an algorithm will be developed in which a network optimization problem is solved by an assignment technique.

Still, both problems are very complicated for real networks in which the enormous number of constraints is a great difficulty. No algorithm seems to exist by which an exact solution may be obtained within a reasonable computation time. Murchland (1969) gives some general principles for devising algorithms. They are partly based on the convex programming theory of Rockafellar (1967) and on some papers by Bruynooghe (1967), Gibert (1968a and b) and Bruynooghe, Gibert and Sakarovitch (1968). Dafermos and Sparrow (1969) propose some algorithms. All these principles and algorithms use at least to a certain degree, the equivalence of the optimization and the equilibrium problem.

For the transportation engineer the equilibrium problem is very common (see also Steenbrink, 1972). It is quite often solved by the so-called capacity restraint technique (see Chicago Area Transportation Study, 1960, Bureau of Public Roads, 1964 and Steel, 1965). In this technique it is attempted to simulate the second principle of Wardrop; so every tripmaker chooses the route with the least users' costs (mostly expressed in travel time), while the costs on a link are an increasing function of the traffic flow on that link. A stepwise assignment technique is used. At each step from all relations of the trip-matrix a part is assigned to the network according to which route has the lowest value for the costs, while these costs are computed with the traffic flows resulting from the preceding step. In practice two to ten steps are used. Often in the later steps smaller parts of the trip-matrix are assigned than in the first step (see also Section 5.3 of this book).

2.1.4.3 Other Aspects of the Assignment

When using the assignment as a simulation of the results of the individual route choice, that is to say as a part of the transport model, other factors come into play.

In the first place, the origin and destination nodes mostly represent centroids of a transport zone. It is assumed that all traffic originates and destinates in those nodes. In reality, however, the trips can be made from any point in a transport zone to any other point in another or in the same transport zone. For this reason alone different paths will in reality be chosen for the same relation (see also Smeed, 1971).

In the second place, we will note the fact that all costs are interpreted individually and so differently by different tripmakers. This also causes the use of different paths for trips in the same relation. This fact is taken into account by the so-called assignment with diversion functions or diversion curves. The diversion function (a kind of demand function) gives the distribution of the trips of the same relation over the different relevant paths. Other factors may play a role, such as the fact that the costs are not known in advance to the tripmaker and so on.

For the most part, in practice, a particular assignment is used, say capacity restraint, which gives a certain distribution over different paths, and it is assumed that it is then no longer necessary to take into account explicitly all the other factors mentioned here.

2.2 DESCRIPTIVE AND NORMATIVE SYSTEMS

2.2.1 Descriptive and Normative Systems

As we have said in the preceding section every trip has its related costs and benefits. Besides the costs and benefits directly borne and received by the tripmaker, there may be costs and benefits borne and received by other members of society.

As costs borne individually by the user (the tripmaker) we may instance his travel time and the fuel he uses etc. The costs to society as a whole include the increase in travel time for the other users of the road, the wear and tear on the road, the damage to the environment etc. From the point of view of the user we may call these the external costs of the trip. As benefits for the user we may instance the benefits received by the traveller by the linking of two activities practised in geographically separate places. As benefits for society as a whole there are the benefits following from the fact that places are visited and that various activities can be carried out at certain places etc.

In the preceding section we have said that we assume that every traveller tries to maximize the difference between his own benefits and costs. We will call the system in which all tripmakers behave according to this individual maximization a *user-optimized or descriptive system*. On the other hand we may think of a system in which all choices are made in such a way that the difference between the benefits and costs for society as a whole is maximized. We will call this system a *society-optimized or normative system*.

The solutions to the user-optimized and the society-optimized system are not necessarily identical. This topic will be further discussed in the following sections. It will be obvious that the difference between the benefits and costs for society will be greatest in the society-optimized system. If the different cost and benefit functions possess certain properties, the solution to the society- and the user-optimized system will coincide.

Of course the difference between the two systems exists in the whole field of transportation. It seems, however, to be best known in the traffic assignment.

2.2.2   Descriptive and Normative Assignment

2.2.2.1   The Difference between the Descriptive and the Normative Assignment

We will assume here that the costs for society consist of the sum of all individual users' costs t^{ab}. The difference between the user-optimized and the society-optimized assignment has been pointed out already by Wardrop (1952). The user-optimized assignment gets the name 'descriptive assignment' because it gives the best description of reality. The society-optimized assignment is called 'normative'. The descriptive assignment equals the equilibrium assignment problem already formulated in Section (2.1.4.1).

To recapitulate, we have the fixed trip-matrix (2.1.21), the network constraints (1.2.11), (1.2.12) and (1.2.13) and the relationships for the costs (1.2.2) (shortest path) and (2.1.6) (increasing costs with increasing traffic flow). The problem is to find the flow pattern such that all tripmakers (users) minimize their own costs:

$$\left. \begin{array}{l} \text{if } x_{ij}^{ab} > 0 \quad \text{then } t^{ai} + t_{ij} = t^{aj} \\[2mm] \text{if } t^{ai} + t_{ij} > t^{aj} \quad \text{then } x_{ij}^{ab} = 0 \end{array} \right\} \quad \begin{array}{l} \text{for all } ab \in P \\[2mm] \text{for all } ij \in L \end{array} \qquad (2.1.5)$$

The normative assignment problem has the same constraints (2.1.21), (1.2.11), (1.2.12), (1.2.13), (1.2.2) and (2.1.6) and is further formulated as the minimization of the total users costs by choice of the flows:

$$\min_{x_{ij}^{P_{ij}^{ab}}} \sum_{ij \in L} F_{ij}(x_{ij}) \qquad (2.2.1)$$

with

$$F_{ij}(x_{ij}) = x_{ij} t_{ij}(x_{ij}) \qquad (2.2.2)$$

As we have seen in the preceding section it is possible to state a minimization problem as an equilibrium problem. The restriction is that the functions $F_{ij}(x_{ij})$ must be convex. With the normal assumption about t_{ij} this is true. Using the same approach as in Section (2.1.3) we see that the minimization problem (2.2.1) is equivalent with the equilibrium problem:

$$\left. \begin{array}{l} \text{If } x_{ij}^{ab} > 0 \quad \text{then } F'^{ai} + F'_{ij} = F'^{aj} \\[2mm] \text{If } F'^{ai} + F'_{ij} > F'^{aj} \quad \text{then } x_{ij}^{ab} = 0 \end{array} \right\} \quad \begin{array}{l} \text{for all } ab \in P \\[2mm] \text{for all } ij \in L \end{array} \qquad (2.2.3)$$

with:

$$F'^{ai} = \sum_{\substack{jk \,\in\, \text{shortest path} \\ \text{from } a \text{ to } i}} \frac{\mathrm{d}F_{jk}}{\mathrm{d}x_{jk}} \qquad (2.2.4)$$

It is obvious now that the results of the normative and the descriptive assignment are generally different. It is more interesting, though, to see when the results of the two assignments are equal. This is obviously the case when:

$$t_{ij} = \frac{\mathrm{d}F_{ij}}{\mathrm{d}x_{ij}} \quad \text{for all } ij \in L \qquad (2.2.5)$$

or because:

$$\frac{\mathrm{d}F_{ij}}{\mathrm{d}x_{ij}} = t_{ij} + x_{ij}\frac{\mathrm{d}t_{ij}}{\mathrm{d}x_{ij}}$$

when:

$$\frac{\mathrm{d}t_{ij}}{\mathrm{d}x_{ij}} = 0 \quad \text{for all } ij \in L \qquad (2.2.6)$$

So the normative and the descriptive assignment have the same result in a network in which the users' costs are not affected by the traffic flows and thus where there is no congestion. This observation seems to have been made first by Jorgensen (1963).

2.2.2.2   The Paradox of Braess

In 1968 Braess published a paradox in transport planning. Murchland (1970) drew attention to it again. This paradox shows very clearly the difference between the normative and the descriptive assignment. The paradox is that the addition of a link to a network can imply an increase in the total users' costs for everybody. Here we give the original example of Braess. The network is shown in Figure 2.2.1.

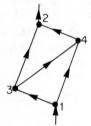

Figure 2.2.1   The network of the paradox of Braess

The following users' costs relationships exist:

$$t_{13} = 10x_{13}; \qquad t_{42} = 10x_{42}$$
$$t_{32} = 50 + x_{32}; \quad t_{14} = 50 + x_{14} \qquad (2.2.7)$$
$$t_{34} = 10 + x_{34}$$

There exists a flow of 6 units from 1 to 2.

In the first situation the link 34 does not exist. The (descriptive) assignment solution is 3 units using the path 132 and 3 units on path 142. The users' costs for each unit are 83. The total users' costs are $6 \times 83 = 498$. In the second situation link 34 is added to the network. When we solve the (descriptive) assignment problem now we get 2 units using path 132, 2 units on path 1342 and also 2 units on path 142. The users' costs for each unit however are now 92; the total users' costs 552. So the addition of a link means an increase of users' costs of about 11 per cent.

In the last situation the assignment with 3 units on path 132 and 3 units on path 142 is not a solution of the assignment problem as formulated in relationship (2.1.5). For in that case it would be better for a unit on 132 with users' costs 83 to use path 1342 with users' costs 81.

We will write down now, for the second situation, the society optimization problem and the equilibrium problem stated as an optimization problem. For the sake of convenience we will use the following notation:

$$x_1 = \text{flow using path 142}$$

$$x_2 = \text{flow using path 1342}$$

$$x_3 = \text{flow using path 132.}$$

The only conservation restriction that remains is then:

$$x_1 + x_2 + x_3 = 6 \tag{2.2.8}$$

Moreover we have the non-negativity restrictions:

$$x_1 \geqslant 0; \quad x_2 \geqslant 0; \quad x_3 \geqslant 0 \tag{2.2.9}$$

We will not introduce the possibly expected integer-constraint (i.e., all x are integers) because, without it, it is possible to use integration and thus to state the descriptive assignment problem as an optimization problem. Moreover it can be shown that the problem has the same solution for $\mathbf{X}$ whether or not it is restricted to being integer.

The society optimization problem becomes:

$$\min_{x_1, x_2, x_3} F = x_1(50 + x_1) + (x_1 + x_2)10(x_1 + x_2) + (x_2 + x_3)10(x_2 + x_3)$$
$$+ x_3(50 + x_3) + x_2(10 + x_2) \tag{2.2.10}$$

subject to restrictions (2.2.8) and (2.2.9). This may be worked out:

$$\min_{x_1, x_2, x_3} F = 11x_1^2 + 21x_2^2 + 11x_3^2 + 20x_1x_2 + 20x_2x_3 + 50x_1$$
$$+ 10x_2 + 50x_3 \tag{2.2.11}$$

subject to restrictions (2.2.8) and (2.2.9).

The equilibrium problem (descriptive assignment) stated as an optimization problem has the form:

$$\min_{x_1, x_2, x_3} F = \int_0^{x_1} (50 + x)\, dx + \int_0^{x_1 + x_2} 10x\, dx + \int_0^{x_2 + x_3} 10x\, dx$$
$$+ \int_0^{x_3} (50 + x)\, dx + \int_0^{x_2} (10 + x)\, dx \tag{2.2.12}$$

subject to restrictions (2.2.8) and (2.2.9). This may be worked out:

$$\min_{x_1, x_2, x_3} F = 11x_1^2 + 21x_2^2 + 11x_3^2 + 20x_1x_2 + 20x_2x_3$$
$$+ 100x_1 + 20x_2 + 100x_3 \tag{2.2.13}$$

subject to restrictions (2.2.8) and (2.2.9).

So we see clearly that problems (2.2.11) and (2.2.13) are two distinct problems and, of course, that the solutions of the two problems will not (always) be the same. Moreover the objective function for the normative assignment problem will never have a larger value than the objective function for the descriptive assignment stated as an optimization problem. Indeed the solution of problem (2.2.13) is $x_1 = 2$, $x_2 = 2$ and $x_3 = 2$ and the solution of problem (2.2.11) is $x_1 = 3$, $x_2 = 0$ and $x_3 = 3$.

2.2.2.3   Reducing the Difference between the Descriptive and the Normative Assignment

It is, of course, beneficial for society as a whole and quite often also for the individual road-user, when there is no difference between the results of the normative and the descriptive assignment. Taking the total users' costs as the total society costs, we have seen that the results of the normative and the descriptive assignment are equal, when the users' costs are independent of the traffic flows.

In other situations the normative and the descriptive assignment do not always give the same solution. To obtain the coincidence of the normative and descriptive assignment again, some extra measures are needed. One possibility is a good (and flexible) traffic management system. Another possibility is the use of the price-mechanism.

Using the price-mechanism means that congestion charges must be imposed. These charges must be so high that the criterion for the individual route choice equals the conditions for the optimal solution of the normative assignment. It is clear from Section 2.2.2.1 that the road-user must regard his costs as $t + x(\mathrm{d}t/\mathrm{d}x)$ instead of t. So the congestion charge must be $x(\mathrm{d}t/\mathrm{d}x)$, which equals the increase of users' costs of all road users. This pricing system, by means of which the users are faced with the marginal costs, is well-known in transport economics (see for instance Oort, 1960).

Indeed, to ensure that everybody is better off, or at any rate that nobody is worse off, it is of course necessary to return the money received in some way (without disturbing the effect of charging it). The practical execution of such a pricing system based on the marginal costs is not a simple task. However, progress seems to have been made in the development of the technical equipment, for instance at the Transport and Road Research Laboratory in England (see also Smeed, 1964).

REFERENCES

Beckmann, M. J., McGuire, C. B., and Winsten, C.B. (1956). *Studies in the Economics of Transportation*, Cowles Commission, Yale University Press, New Haven.

Bergendahl, G. (1969). *Models for Investments in a Road Network*, Bonniers, Stockholm.

Braess, D. (1968). Über ein Paradoxen der Verkehrsplanung, *Unternehmensforschung*, **12**, 258–268.

Bruynooghe, M. (1967). *Affectation du Trafic sur un Multi-réseau*, Institut de Recherche des Transport, Arcueil.

Bruynooghe, M., Gibert, A., and Sakarovitch, M. (1968). Une méthode d'affectation du trafic. Paper presented at the Fourth International Symposium on *The Theory of Traffic Flow*, Karlsruhe (June).

Bureau of Public Roads (1964) *Traffic Assignment Manual*, United States Government Printing Office, Washington D.C.

Charnes, A., and Cooper, W. W. (1966). *Simulation, Optimization and Evaluation of Systems of Traffic Networks*, Management Science Research Report, No. 77, Carnegie Institute of Technology, Pittsburgh. Penn.

Chicago Area Transportation Study (1960). *Final Report*, Chicago.

Colony, D. C. (1970). An application of game theory to route selection. *Traffic Flow Theory, 6 Reports*, Highway Research Record, No. 334.

Dafermos, S. C., and Sparrow, F. T. (1969). The traffic assignment problem for a general network, *Journal of Research of the National Bureau of Standards; B. Mathematical Sciences*, **73B**, No. 2 (April–June).

Gibert, A. (1968a). *A Method for the Traffic Assignment Problem when Demand is Elastic*, Transport Network Theory Unit Report LBS-TNT 85, London Business School, London.

Gibert, A. (1968b). *A method for the Traffic Assignment Problem*, Transport Network Theory Unit Report LBS-TNT 95, London Business School, London.

Hamerslag, R., in collaboration with Steenbrink, P. A. (1970). Het voorspellen van vervoersstromen, *Verkeerstechniek*, **21**, Nos. 5, 6 and 7 (May, June, July).

Hamerslag, R. (1972). *Prognosemodel voor het Personenvervoer in Nederland*, Koninklijke Nederlandse Toeristenbond A.N.W.B., 's-Gravenhage.

Henderson, J. M., and Quandt, R. E. (1958). *Microeconomic Theory: a Mathematical Approach*, McGraw-Hill, New York.

Jorgensen, N. O. (1963). *Some aspects of the urban traffic assignment problem*, Graduate Report, Institute of Transportation and Traffic Engineering, University of California, Berkeley.

Kohl, J. E. (1841). *Der Verkehr und die Ansiedelung der Menschen in ihrer Abhängigkeit von der Gestaltung der Erdoberfläche*, Dresden, Leipzig.

Kuhn, H. W., and Tucker, A. W. (1951). Nonlinear programming. *Proceedings of the Second Berkeley Symposium on Mathematical Statistics and Probability*, Neyman, J. ed., University of California Press.

Murchland, J. D. (1969). *Gleichgewichtsverteilung des Verkehrs im Strassennetz*, Verlag Anton Hain, Meisenheim. Also in English (1969): *Road Network Traffic Distribution in Equilibrium*. Paper presented at the conference: 'Mathematical Methods in the Economic Science'. Oberwolfach.

Murchland, J. D. (1970). *Braess's Paradox of Traffic Flow*, Institut für Angewandte Reaktorphysik, Karlsruhe; also published in *Transportation Research*, **4**.

Neumann, J. von, and Morgenstern, O. (1947). *Theory of Games and Economic Behavior*, Princeton University Press, Princeton N.J.

Oort, C. J. (1960). *Het Marginalisme als Basis voorde Prijsvorming in het Vervoerswezen*; *een Analyse*. Stichting Verkeerswetenschappelijk Centrum, Rotterdam.

Overgaard, K. R. (1966). *Traffic Estimation in Urban Transportation Planning*, Acta Polytechnica Scandinavia.

Rockafellar, R. T. (1967). Convex programming and systems of elementary monotonic relations. *Journal of Mathematical Analysis and Applications*, **19**.

Smeed, R. J., and others (1964). *Road Pricing: The Economic and Technical Possibilities*. Her Majesty's Stationery Office, London.

Smeed, R. J. (1971). *Effect of Zone Size on Assignment*. Research Report Research Group in Traffic Studies, University College, London.

Steel, M. A. (1965). Capacity restraint, a new technique. *Traffic Engineering and Control* (October).

Steenbrink, P. A. (1972). Netwerktoedeling, optimalisering en invoerverzorging, *Verkeerstechniek*, **23**, No. 11 (November).

Timman, R. (1966). *Optimaliseren van Funkties en Funktionalen*, Technische Hogeschool, Delft, Onderafdeling der Wiskunde, Delft.

Wardrop, J. G. (1952). Some theoretical aspects of road traffic research. *Proceedings, Institute of Civil Engineers*, **1**, 325–362.

Wilson, A. G. (1967). A statistical theory of spatial distribution systems. *Transportation Research*, **1**, 253–269.

Wilson, A. G., and Wagon, D. J. (1968). *The SELNEC Transport Model*. Mathematical Advisory Unit Note 200, Department of the Environment, London.

Wohl, M., and Martin, B. V. (1969). *Traffic Systems Analysis for Engineers and Planners*, McGraw-Hill, New York.

3

The Transport Network Optimization Problem

In this chapter we will state in general terms the transport network optimization problem. That means that we will present possibilities for the objective function, the decision variables and the constraints, all defined on a transport network. General surveys of the transport network optimization problem have also been given for instance by Manheim *et al.* (1968) and by Bergendahl (1969). Extensive lists of references for this problem can be found in Bhatt (1968) and in Jansen (1971).

3.1 OPTIMAL PHYSICAL PLANNING

In Section 2.1.1 transportation has been defined as consisting of the transfer of persons and/or goods between geographically separate places. Moreover in Section 2.1.2 the benefits of a trip have been said to arise from the linking of two activities separated geographically. So it is obvious that there will exist a close relationship between transportation and the spatial structure. Firms and people settle in places, that are easily accessible, i.e. near roads and railways. On the other hand the need for transportation and its infrastructure is highest where many people and firms have settled. Problems such as the function of the inner part of a town or the desirability of spatial concentration or deconcentration are problems of physical planning as well as transportation planning. It is often argued that the construction or improvement of transport infrastructure is one of the most important instruments for the implementation of a regional economic development policy, along with geographically specified subsidies and/or restrictions for settlements and improvements of the social infrastructure (education, cultural provisions) in a special region (see for instance van de Poll and Bourdrez, 1972). At any rate it is clear that we must treat the system of physical and transportation planning as one system and that we must try to find the optimal solution for this system. But it will be clear too that the finding of this *optimum optimorum* is not an easy task.

The first, but probably also the most difficult, thing to do is to define the objective function, the decision variables and the constraints. To narrow down the possibilities for the objective function we can try to split the effects of some spatial structure into benefits and costs for the society. The difference between these benefits and costs might then be chosen as the objective function to be maximized.

For the benefits we look at the contribution of the spatial structure to the well-being of the society. There is the contribution it makes to more-or-less well-defined economic variables such as the level of the national income; and it also affects the geographical and interpersonal distribution of this national income. In addition, the quality of social life is raised by the possibilities for communication. The equilibrium and the diversity of the whole ecosystem is also involved. On the costs side we should take into account the use of land, raw materials and labour for the construction and maintenance of places for living and working, and the costs of transportation.

It is essential that all elements of the objective function are given as quantitative variables, expressed in the same units. Therefore it is necessary to have available a good system of indicators for the well-being of a town or a nation. This is, of course, a very difficult problem and one which seems closer to being solved in economics (the National Accounts, for instance) than in social sciences (see for instance Bauer, 1966, Koelle, 1968 and Chapter 14 of Steenbrink, 1969). Tinbergen (1957) for instance considers the effect of road construction on some economic variables. Ben-Shahar et al. (1969) state the town-planning problem as a linear-programming model. They use a simple but worked-out objective function including economic and social variables (see also van Est, 1972).

The decision variables for the problem of optimal physical planning might include figures for population, working population and employment in each industry for every area, the transportation systems and so on.

There is also a great deal of literature on problems stated more theoretically, e.g. the ideal spatial structure for a homogeneous group of people. These problems and their solutions are often based on geometrics. See for instance Klaassen (1969) and Domańsky (1967).

In this study we will not deal with the problem of optimal physical planning. Instead we will adopt the approach of trying to find the optimal transport infrastructure for a given spatial distribution of population and employment.

It is, of course, possible to state and solve this transport network optimization problem for a number of spatial distributions and to compare the different results. The problem of choice and evaluation still remains. A factor to be considered is that the (optimized) transport infrastructure should be consistent with given land-use patterns. In other words: Is the quality of the transport infrastructure such that people are living and working where they are supposed to be living and working according to the given land-use pattern?

3.2 OPTIMAL TRANSPORT INFRASTRUCTURE FOR A GIVEN LAND USE

3.2.1 The Decision Variables

We will presume that the decision variables for the optimization problem will be set up by one body, a central authority, for instance, trying to regulate the transportation system. The choices of destination, travel mode and route can also be decision variables. But, in that case, we presume that these variables are fixed by the central authority, so a normative system is used. The descriptive system, in which the different decisions are presumed to be taken in reality by the individual tripmakers, will be taken into account as a set of constraints for the transport network optimization problem. We presume a dynamic situation in which the transport network can be changed.

There are now two main classes of 'real' instrument variables for the regulating authority. One class is related to the transport networks themselves, the other class consists of instruments that can be used to influence the use of the networks.

To the first class we ascribe the (changes in the) transport networks. The networks have their structure and their dimensions. For public transport the operation schemes, the number and location of stations and stops and so on are also important. Parking facilities too are instrument variables.

One of the most important instruments that can be used to influence the use of transportation systems is the pricing system. We may think here of tolls for road use, fares in public transport, fuel taxes, taxes on the purchase and possession of vehicles and so on. Apart from these, the traffic management system is also an instrument variable. In this study we will only deal with the first class of instruments. For the pricing system used as an instrument, see for instance Oort (1960), Smeed (1964) and de Donnea (1971).

There is one further remark to be made with respect to the transport network as an instrument variable. Because a transport network is already in existence changes in the existing network are passed on. These changes are accomplished by (some) investment. One may also speak of costs and benefits with regard to this investment. The costs are the construction costs etc., and the decrease in other costs in the system transportation, e.g. users' costs, are the benefits.

3.2.2 The Objective Function

3.2.2.1 General Remarks

Before starting a discussion on the objective function we must emphasize primarily that the defining of the objective function is not the task of the researcher. So, in this section, only some possibilities for the objective function will be mentioned. The ultimate evaluation and choice will be left to the user of the optimization system.

Some people state that it is impossible to define a reasonable objective function for the transport network optimization problem, in which all relevant factors are included completely and consistently. The reason put forward for this impossibility is that it is impossible to evaluate all relevant factors in the same units, say monetary units, and that there will never be full agreement about the objectives among all decision makers. To overcome this difficulty Manheim and Hall (1968) propose, for instance, to work with a 'goal-fabric'. All factors to be taken into account are made into a list of goals, which needs not necessarily to be complete or consistent. Many optimization problems are now stated with an objective function composed of some or all of the goals, weighted in a certain way. An interesting remark made by Lindblom on this point (1965), is that sometimes the same values for the decision variables can be chosen to optimize different objective functions.

A plausible and reasonable objective may be the maximization of the (positive) difference between the benefits and costs of the transportation system to the society. We will henceforth call this the *social surplus* or '*surplus*' for short.

In Section 2.1.2 we have said already that the social benefits consist of benefits to the tripmaker (linking of two activities) and benefits to others (the effect on the spatial distribution, the benefits arising from the possibility of being visited, and so on.) For the last type of benefit the word 'accessibility' is sometimes used.

We can also divide the social costs into different types of costs:

(a) costs directly related to the trip and generally borne by the tripmakers (so called users' costs as traveltime consumption, fuel costs etc.);

(b) costs related to the construction and maintenance of the infrastructure networks; we include the exploitation costs of public transport in this type; these costs are generally borne by the regulating authority;

(c) costs, related to the infrastructure or the use of it, but generally borne by 'third parties' (e.g. neighbours); these costs are often called external costs; the damage to the environment is an example.

It is very difficult to define, to measure and to evaluate the social costs. One must also be very careful to get no double-counting (including, for instance, both fares paid on public transport and the exploitation costs of public transport). Although the defining and evaluation of the social costs is very difficult, it seems to be possible to get a fairly workable estimate for them. For a further discussion of this subject see Chapter 9 in the second part of this book.

Even more difficult is the problem of defining, measuring and evaluating the social benefits. This problem does not seem to have been solved. It is clear that this problem is closely related to the evaluation of land use, a problem we mentioned in the preceding section. Some authors use land prices to estimate the effect of infrastructure on land use (e.g. Mohring and Harwitz, 1962; Gwilliam, 1970; Ben-Shahar et al., 1969). Land prices reflect two effects of infrastructure: prices increase near a new road; this is the result of the increase of accessibility, a benefit; but prices also fall near a new road; this is the result of the 'external' costs as noise and so on.

In defining the total benefits for the users, or, to put it a better way, the changes in the total users' benefits, we seem to be more successful. The keywords for the methods used here are 'consumers' surplus' and 'willing-ness-to-pay'. We will discuss the concept of consumers' surplus and its use in transportation planning in the next subsection. Although this consumers' surplus theory seems to give a very valuable and practical tool for estimating changes in users' benefits, many authors have objections to its use.

There is another way of stating the objective function which avoids the difficult problem of evaluating the benefits to users and society: we assume the benefits to be the fact that people can travel and do travel between the different places; thus we assume the trip-matrix to be the benefits; keeping these benefits, and thus the trip-matrix, constant, the maximization of the difference between benefits and costs boils down to the minimization of the costs.

3.2.2.2 Consumers' Surplus in Transportation Planning

The concept of consumers' surplus dates back to Dupuit (1844) and was popularized by Marshall (1930). A recent survey is given by Currie, Murphy and Schmitz (1971). The use of this concept in transportation planning for network optimization or evaluation is described by, among others, Mohring and Harwitz (1962), Quarmby (1969), Neuburger (1969 and 1971), Bergen-dahl (1969) and Smith (1970).

To derive the basic principles of this consumers' surplus theory in trans-port planning, we return to what we said in Section 2.1.2 about the behaviour of tripmakers. It is presumed that these tripmakers try to maximize the difference between their benefits and their costs.

Let us start with the simple case of one road and one relation, ab. According to relationship (2.1.1) the following relationship will be satisfied if a tripmaker decides to travel from a to b:

$$u^{ab} - t^{ab} \geqslant u^{aa} \qquad (3.2.1)$$

with:

u^{aa}—benefits for the tripmaker of staying in a

$t^{aa} = 0$ (travel costs of staying in a are omitted)

As we know, the benefits and costs are individual variables. So for some people a change in the costs will mean a change in their behaviour and for other people it will not. Let us focus on a person who does change. Let us suppose that he does travel from a to b when the costs are t^{ab}. So relationship (3.2.1) is satisfied. When the costs, however, are $t^{ab} + \Delta t^{ab}$ we assume that he does not travel any more. In that case relationship (3.2.2) is satisfied:

$$u^{ab} - (t^{ab} + \Delta t^{ab}) \leqslant u^{aa} \qquad (3.2.2)$$

When Δt^{ab} approaches zero (3.2.3) holds for this person:

$$u^{ab} - u^{aa} = t^{ab} \qquad (3.2.3)$$

The cost t^{ab} is the 'price' which that particular person is willing to pay to travel from a to b. Therefore t^{ab} is called the person's willingness-to-pay. When the actual costs to the tripmaker (consumer) are t'^{ab} (with $t'^{ab} \leqslant t^{ab}$) it is said that the tripmaker enjoys a consumers' surplus of $t^{ab} - t'^{ab}$.

We may define the total users' benefits of travelling from a to b as the summation of all individual users' benefits. Using the inverse of the (collective) demand function [(2.1.4): $t^{ab} = g^{ab}(x^{ab})$] we get:

$$U^{ab} - U^{aa} = \int_{0}^{x^{ab}} g^{ab}(x)\,\mathrm{d}x \qquad (3.2.4)$$

with:

U^{ab}—total users' benefits of travelling from a to b for the tripmakers x^{ab}.

U^{aa}—total users' benefits of staying in a for the tripmakers x^{ab}.

There arise some practical difficulties in defining the total users' benefits as has been done in relationship (3.2.4):

(a) the (inverse) demand function is mostly only known on a small interval of x:

(b) the (inverse) demand function is anyway not defined for $x = 0$.

These difficulties are solved by defining not the total users' benefits, but only the change in these benefits. To do that we assume a change from the situation I with x_I^{ab} and t_I^{ab} to the situation II with x_{II}^{ab} and t_{II}^{ab}. The total change in benefits then becomes:

$$\Delta(U^{ab} - U^{aa}) = (U^{ab} - U^{aa})_{II} - (U^{ab} - U^{aa})_{I} = \int_{x_I^{ab}}^{x_{II}^{ab}} g^{ab}(x)\,\mathrm{d}x \quad (3.2.5)$$

This last value is often approximated by the area of a trapezium (see Figure 3.2.1):

$$(U^{ab} - U^{aa}) = \tfrac{1}{2}(x_{II}^{ab} - x_I^{ab})(t_{II}^{ab} + t_I^{ab}) \qquad (3.2.6)$$

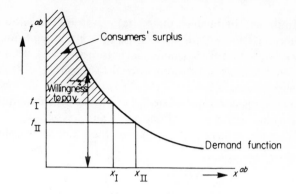

Figure 3.2.1   Consumers' surplus

Besides the change in the users' benefits the change in the users' costs and the change in the difference of these two variables (i.e., the change in the consumers' surplus) is also of interest.

Still working with one relation, ab, and one road and the situation I before and II after, the total change in the consumers' surplus becomes:

$$\Delta S = \Delta(U - T) = \Delta U - \Delta T = \int_{x_I^{ab}}^{x_{II}^{ab}} g^{ab}(x)\, dx - (x_{II}^{ab} t_{II}^{ab} - x_I^{ab} t_I^{ab}) \qquad (3.2.7)$$

with S–total consumers' surplus for travelling from a to b.

Again approximating the integral by the area of a trapezium we get:

$$\Delta S = \tfrac{1}{2}(x_{II}^{ab} - x_I^{ab})(t_{II}^{ab} + t_I^{ab}) - (x_{II}^{ab} t_{II}^{ab} - x_I^{ab} t_I^{ab}) = \tfrac{1}{2}(x_I^{ab} + x_{II}^{ab})(t_I^{ab} - t_{II}^{ab})$$
$$(3.2.8)$$

It is possible to derive the results for the value of the total change in consumers' surplus (3.2.7) and (3.2.8) in another way. For this we split the total group of (potential) tripmakers into three subgroups:

(a)   the people travelling in the situations I and II;
(b)   the people not travelling in situation I and travelling in situation II;
(c)   the people travelling neither in situation I nor in II.

We assume that $x_I^{ab} < x_{II}^{ab}$ and $t_I^{ab} > t_{II}^{ab}$. It is easy to give a similar derivation for the reverse situation.

For the subgroup (a) there is no change in benefits and the following change in costs:

$$\Delta T_a = x_I^{ab}(t_{II}^{ab} - t_I^{ab})$$

So the change in surplus is:

$$\Delta S_a = x_I^{ab}(t_I^{ab} - t_{II}^{ab}) \qquad (3.2.9)$$

The subgroup (b) can be further split by taking all individuals deciding to travel from a to b if $t^{ab} = t'^{ab}$ for all t'^{ab} with $t_{II}^{ab} \leqslant t'^{ab} \leqslant t_{I}^{ab}$. From relationship (3.2.3) we know that if somebody starts to travel when $t^{ab} = t'^{ab}$, that, for that person, $u^{ab} - u^{aa} = t'^{ab}$ holds. So, before travelling, that person's consumers' surplus was $u^{aa} - t^{aa} = u^{aa}$. At the point $t^{ab} = t'^{ab}$ when he is actually travelling his consumers' surplus is $u^{ab} - t'^{ab}$. So the difference in consumers' surplus between the situations I and $'$ is $u^{ab} - t'^{ab} - u^{aa} = 0$. If the costs decrease further to t_{II}^{ab} the change in consumers' surplus is given by the change in costs:

$$\Delta s = -(t_{II}^{ab} - t'^{ab}) \tag{3.2.10}$$

Using the inverse of the demand function, the total value of the consumers' surplus of the subgroup (b) can be written as:

$$\Delta S_b = -\int_{x_{I}^{ab}}^{x_{II}^{ab}} (t_{II}^{ab} - g^{ab}(x))\, dx = -t_{II}^{ab}(x_{II}^{ab} - x_{I}^{ab}) + \int_{x_{I}^{ab}}^{x_{II}^{ab}} g^{ab}(x)\, dx \tag{3.2.11}$$

Of course it is possible to approximate equation (3.2.11) by the area of a (very well-known) triangle:

$$\Delta S_b = \tfrac{1}{2}(x_{II}^{ab} - x_{I}^{ab})(t_{I}^{ab} - t_{II}^{ab}) \tag{3.2.12}$$

For the subgroup (c) there is a change neither in benefits nor in costs and, therefore, no change in surplus:

$$\Delta S_c = 0 \tag{3.2.13}$$

Adding equations (3.2.9), (3.2.12) and (3.2.13) we get the total change in consumers' surplus:

$$\Delta S = \tfrac{1}{2}(x_{I}^{ab} + x_{II}^{ab})(t_{I}^{ab} - t_{II}^{ab})$$

which equals the formerly derived result (3.2.8).

Of course, it is also possible to state upper and lower bounds for the integral and so for the total change in consumers' surplus (again we suppose $t_{I}^{ab} \geqslant t_{II}^{ab}$ and $x_{I}^{ab} \leqslant x_{II}^{ab}$; the same relationships can be derived for the reverse situation):

$$\left.\begin{aligned} \sup(\Delta U) &= t_{I}^{ab}(x_{II}^{ab} - x_{I}^{ab}) \\ \inf(\Delta U) &= t_{II}^{ab}(x_{II}^{ab} - x_{I}^{ab}) \end{aligned}\right\} \tag{3.2.14}$$

$$\left.\begin{aligned} \sup(\Delta S) &= x_{II}^{ab}(t_{I}^{ab} - t_{II}^{ab}) \\ \inf(\Delta S) &= x_{I}^{ab}(t_{I}^{ab} - t_{II}^{ab}) \end{aligned}\right\} \tag{3.2.15}$$

(sup (=supremum) indicates the least upper bound, inf (=infimum) the greatest lower bound).

Till now we only have dealt with the consumers' benefits and surplus for the simple situation of one relation and one road. We will turn now to the more complicated situation of many relations and a whole network. The simple demand function does not hold any longer. The demand for transport in relation ab is now also dependent on the costs and benefits of travelling in all other relations. We will now give a similar analysis as for the simple case.

If at first somebody travels in relation ab and, after a change Δt^{ab} in costs, he travels in relation ac, then we know that for that person (3.2.16) holds:

$$\left.\begin{array}{c} u^{ab} - t^{ab} \geqslant u^{ac} - t^{ac} \\ u^{ab} - (t^{ab} + \Delta t^{ab}) \leqslant u^{ac} - t^{ac} \end{array}\right\} \qquad (3.2.16)$$

Letting Δt^{ab} approach zero we get:

$$u^{ab} - t^{ab} = u^{ac} - t^{ac} \qquad (3.2.17)$$

So at the moment of changing from one relation to another the consumers' surplus is the same for both relations.

Focusing on one relation we can derive the change in the total consumers' surplus obtained by travelling in that relation. We use the fact that when people do not change their travel relations, there is no difference in consumers' benefits and that at the moment of changing there is no difference in consumers' surplus.

Because in the many relations/network case simple demand functions like relationship (2.1.4) do not exist, we cannot use integration to derive the relationships for the changes in total consumers' surplus. Instead we set upper and lower bounds on those changes and get the following relationships:

$$\left.\begin{array}{l} \text{if } x_{\text{II}}^{ab} \geqslant x_{\text{I}}^{ab} \quad \text{and} \quad t_{\text{II}}^{ab} \leqslant t_{\text{I}}^{ab}: \\[4pt] \qquad \sup \Delta S^{ab} = x_{\text{II}}^{ab}(t_{\text{I}}^{ab} - t_{\text{II}}^{ab}) \\[4pt] \qquad \inf \Delta S^{ab} = x_{\text{I}}^{ab}(t_{\text{I}}^{ab} - t_{\text{II}}^{ab}) \\[6pt] \text{if } x_{\text{II}}^{ab} \leqslant x_{\text{I}}^{ab} \quad \text{and} \quad t_{\text{II}}^{ab} \geqslant t_{\text{I}}^{ab}: \\[4pt] \qquad \sup \Delta S^{ab} = x_{\text{II}}^{ab}(t_{\text{I}}^{ab} - t_{\text{II}}^{ab}) \\[4pt] \qquad \inf \Delta S^{ab} = x_{\text{I}}^{ab}(t_{\text{I}}^{ab} - t_{\text{II}}^{ab}) \\[6pt] \text{if } x_{\text{II}}^{ab} \geqslant x_{\text{I}}^{ab} \quad \text{and} \quad t_{\text{II}}^{ab} \geqslant t_{\text{I}}^{ab}: \\[4pt] \qquad \sup \Delta S^{ab} = x_{\text{I}}^{ab}(t_{\text{I}}^{ab} - t_{\text{II}}^{ab}) \\[4pt] \qquad \inf \Delta S^{ab} = x_{\text{II}}^{ab}(t_{\text{I}}^{ab} - t_{\text{II}}^{ab}) \\[6pt] \text{if } x_{\text{II}}^{ab} \leqslant x_{\text{I}}^{ab} \quad \text{and} \quad t_{\text{II}}^{ab} \leqslant t_{\text{I}}^{ab}: \\[4pt] \qquad \sup \Delta S^{ab} = x_{\text{I}}^{ab}(t_{\text{I}}^{ab} - t_{\text{II}}^{ab}) \\[4pt] \qquad \inf \Delta S^{ab} = x_{\text{II}}^{ab}(t_{\text{I}}^{ab} - t_{\text{II}}^{ab}) \end{array}\right\} \qquad (3.2.18)$$

Here, too, it is of course possible to approximate the relationships (3.2.18) by taking the area of a trapezium:

$$\Delta \tilde{S}^{ab} = \tfrac{1}{2}(x_I^{ab} + x_{II}^{ab})(t_I^{ab} - t_{II}^{ab}) \qquad (3.2.19)$$

To get the total change in consumers' surplus it is necessary to sum over all relations:

$$\left. \begin{aligned} \sup \Delta S &= \sum_{ab \in P} \sup \Delta S^{ab} \\ \inf \Delta S &= \sum_{ab \in P} \inf \Delta S^{ab} \\ \Delta \tilde{S} &= \sum_{ab \in P} \Delta \tilde{S}^{ab} \end{aligned} \right\} \qquad (3.2.20)$$

The difference between the upper and the lower bound may give some idea how close the approximation by the trapezium-area is or how 'sure' the beneficial or non-beneficial effects of the improvements in the transport networks are.

Note that the summation is over all relations and not over all links. The latter value is generally different from the former and gives a wrong value for the change in consumers' surplus.

As said before, many authors use this consumers' surplus theory to obtain an estimate for the change in the users' benefits and costs. Smith (1970) and Neuburger (1971) even use it, though only under certain conditions, to give an evaluation for different land-use schemes.

Although the consumers' surplus theory seems to give a very valuable practical tool to estimate changes in users' benefits, many authors have objections to its use (see for instance: Henderson and Quandt (1958), Chapter 7, *Welfare Economics*, Oort (1966) and Samuelson (1967)). Major objections are that (1) individual benefits have been summed, a procedure which cannot be considered correct; (2) the distribution of benefits is left out of consideration: and (3) surpluses in other sectors, appearing or disappearing through changes in the demand for transport, are not taken into account.

3.2.3 The Constraints

Network constraints on a transport network optimization problem will always be present. They consist of the conservation laws (relationship (1.2.11)), the non-negativity restrictions (1.2.12) and the summations of the flows due to the different relations on a link (relationship (1.2.13)).

Whether the following sets of constraints apply depends on the statement of the problem. For the case of costs minimization the fixed trip-matrix is an important constraint:

$$x^{ab} = x^{0\,ab} \quad \text{for all } ab \in P \qquad (3.2.21)$$

For the case of surplus maximization no such constraint exists.

Another important distinction is to be made between descriptive and normative systems. For the descriptive systems the whole set of transportation models forms an important set of constraints. In this book we will usually represent these constraints as:

$$G(X, C) = 0 \tag{3.2.22}$$

For the normative systems no such constraints exist.

Besides these fundamental constraints, there are many other possible constraints. We will represent these additional constraints usually as:

$$H(X, C) >, =, < 0 \tag{3.2.23}$$

Below we will just mention some cases illustrating possible additional constraints:

(a) the total investments are not allowed to exceed a certain budget; and this budget can be fixed per region R or per time period t:

$$\sum_{ij \in L} i_{ij}(c_{ij}) \leqslant I^0 \tag{3.2.24}$$

$$\sum_{ij \in L_R} i_{ij}(c_{ij}) \leqslant I_R^0 \tag{3.2.25}$$

$$\sum_{ij \in L} i_{tij}(c_{tij}) \leqslant I_t^0 \tag{3.2.26}$$

(b) a certain level of effectiveness of the transportation system is required; this means that the users' costs are not allowed to exceed a certain value; the total users' costs may be involved in this, or the users' costs for every relation or just the level of service for every road:

$$\sum_{ab \in L} t^{ab} \leqslant T^0 \tag{3.2.27}$$

$$t^{ab} \leqslant l^{ab} t^0 \quad \text{for all } ab \in P \tag{3.2.28}$$

$$t_{ij} \leqslant l_{ij} t^0 \quad \text{for all } ij \in L \tag{3.2.29}$$

(c) for reasons of safety, or for other reasons, a minimum value for the users' costs (say maximum speed) can be required:

$$t_{ij} \geqslant l_{ij} t^1 \quad \text{for all } ij \in L \tag{3.2.30}$$

(d) the dimensions of the different roads in the networks may be limited to certain values:

$$c_{ij}^{\min} \leqslant c_{ij} \leqslant c_{ij}^{\max} \quad \text{for all or some } ij \in L \tag{3.2.31}$$

(e) concern for the employment situation of the whole study area or of some regions may be demanded; or considerations of regional income distribution may have to be taken into account:

$$\sum_{ij \in L} i_{ij}(c_{ij}) \geqslant I^1 \tag{3.2.32}$$

$$\sum_{ij \in L_R} i_{ij}(c_{ij}) \geqslant I_R^1 \tag{3.2.33}$$

(f) only one of two possible alternative roads can be constructed:

$$c_{ij} c_{kl} = 0 \quad (ij \neq kl) \tag{3.2.34}$$

(g) it may be necessary to follow a particular sequence in construction:

$$c_{t+1\,ij} \geqslant c_{tij} \tag{3.2.35}$$

or

$$c_{t+1\,ij} \geqslant c_{tkl} \quad (ij \neq kl) \tag{3.2.36}$$

It is of course possible to extend this list indefinitely with other possible constraints.

There is one more category of constraints to be considered. These occur only when the problem is stated in a specific way. Well-known specific statements of the problem are: the minimization of the investment costs for a given level of effectiveness (relationships (3.2.27) up to and including (3.2.29)), or the minimization of users' costs for a fixed budget for investments (3.2.24).

REFERENCES

Bauer, R. A. (1966). *Social Indicators*, M.I.T. Press, Cambridge, Mass.

Ben-Shahar, H., Mazor, A., and Pines, D. (1969). Town planning and welfare maximization: a methodological approach, *Regional Studies*, 3, 105–113.

Bergendahl, G. (1969). *Models for Investments in a Road Network*, Bonniers, Stockholm.

Bhatt, K. U. (1968). *Search in Transportation Planning: A Critical Bibliography*, Research Report R 68-46, M.I.T. Department of Civil Engineering, Cambridge, Mass.

Currie, J. M., Murphy, J. A., and Schmitz, A. (1971). The concept of economic surplus and its use in economic analysis. *The Economic Journal* (December)

Domański, R. (1969). Remarks on simultaneous and anisotropic models of transportation network. *The Regional Science Association Papers*, 19, 223–228.

Donnea, F. X. de (1971). *Richtlijnen voor het Prijsmechanisme voor het Weggebruik in Nederland*, Deelrapport 24, Integrale Verkeers- en Vervoerstudie door het Nederlands Economisch Instituut, Rotterdam.

Dupuit, J. (1844). On the measurement of the utility of public works. *Annales des Ponts et Chaussées*, Deuxième Serie, Vol. 8; translation reprinted in Munby, D. (1968), *Transport*.

Est, J. P. J. M. van (1972). *Environmental Planning and Welfare Maximization*, Internal report of Adviesburo voor Verkeersordening, Goudappel & Coffeng, Deventer.

Gwilliam, K. M. (1970). The indirect effects of highway investment, *Regional Studies*, 4, No. 2 (August).

Ienderson, J. M., and Quandt, R. E. (1958). *Microeconomic Theory: a Mathematical Approach*, McGraw-Hill, New York.

ansen, G. R. M. (1971). *Transport Network Optimization: a Preliminary Bibliography*, 2nd edition. Working Paper No. OTN/3/71.4, Transportation Research Laboratory, Delft University of Technology, Delft.

laassen, L. H. (1969). *The Role of Traffic in the Physical Planning of Urban Areas*, Report prepared for the CEMT Symposium, Rome (September).

oelle, H. H. (1968). Sozio-ökonomisches Modell des Planeten Erde, Arch+, 1, No. 1.

indblom, C. E. (1965). *The Intelligence of Democracy*, The Free Press, New York.

Ianheim, M. L., and Hall, F. L. (1968). *Abstract Representation of Goals*, Professional Paper P67-24, M.I.T. Department of Civil Engineering, Cambridge, Mass. also in (1968): *Transportation: A Service, Proceedings of the New York Academy of Sciences*, American Society of Mechanical Engineers, Transportation Engineering Symposium.

Ianheim, M. L., with Ruiter, E. R. and Bhatt, K. U. (1968). *Search and Choice in Transport Systems Planning: Summary Report*, Research Report R68-40; M.I.T. Department of Civil Engineering, Cambridge, Mass.

Iarshall, A. (1930). *Principles of Economics*, London.

Iohring, H. and Harwitz, M. (1962). *Highway Benefits*, Northwestern University Press, Evanston, Ill.

Ieuburger, H. (1969). *The evaluation of User Benefits on Transport Projects*, Economic Planning Directorate Technical Note 2, Ministry of Transport, London.

Ieuburger, H. (1971). User benefit in the evaluation of transport and land use plans. *Journal of Transport Economics and Policy*, 5, No. 1 (January).

Iort, C. J. (1960). *Het Marginalisme als Basis voorde Prijsvorming in het Vervoerswezen; een Analyse*, Stichting Verkeerswetenschappelijk Centrum, Rotterdam.

Iort, C. J. (1966). *De Infrastructuur van het Vervoer*, Algemene Verladers en Eigen Vervoerders Organisatie E.V.O., 's-Gravenhage.

oll, E. H. van de en Bourdrez, J. A. (1972). Infrastructuur en regionale ontwikkeling. In Klaassen, L. H. (ed.), *Regionale Economie, Het Ruimtelijk Element in de Economie*, Wolters-Noordhoff, Groningen.

Iuarmby, D. A. (1969). *An Interim Procedure for the Economic Evaluation of Transportation Networks*, Mathematical Advisory Unit Note 136, Ministry of Transport, London.

amuelson, P. A. (1967). *Foundations of Economic Analysis*, New York.

meed, R. J., and others (1964). *Road Pricing: The Economic and Technical Possibilities*, Her Majesty's Stationery Office, London.

mith, J. F. (1970). The design of a transportation study with regard to the evaluation of its output. *Regional Studies*, 4, No. 2 (August).

teenbrink, P. A. (ed.) (1969). *Future Game—Een Macro-economisch Beleidsspel*, Universitaire Pers Rotterdam, Wolters–Noordhoff, Groningen.

inbergen, J. (1957). The appraisal of road construction: two calculation schemes, *The Review of Economics and Statistics* (August).

4

Known Solution Methods for the Optimization of Transport Networks

In this chapter we will discuss the most important solution methods to be found in the literature for the transport network optimization problem. Some new extensions will also be given (Sections 4.3.6 and 4.4.2). A general survey of possible solution methods has been given, for instance, by Manheim *et al.* (1968) and by Bergendahl (1969). Extensive lists of references are provided by Bhatt (1968) and by Jansen (1971).

4.1 SHORT STATEMENT OF THE PROBLEM

In this first section we will give a clear statement of the problem, repeating in brief part of Chapter 3. We will focus on just one trip-matrix. So we are looking at the static situation (no changes over time) and at one transport mode, ignoring the fact that the traffic is not constant during the hours of the day or over the days of the year.

In Chapter 2 we introduced the normative and the descriptive system. We will deal here with both. So the traffic flows will in some cases, and will not in other cases, be decision variables. The most important decision variable we will use (Section 3.2.1) is the transport network. We will consider the (maximal possible) structure as given and focus on the choice of the dimensions of the links in the network. But we will include zero as a possible dimension. It is in fact thus that a particular structure is chosen from a set of possible structures.

For the objective function we will use the social surplus S, (Section 3.2.2). the difference between the social benefits U and the social costs F. For reasons of simplicity we assume that the social costs are composed of cost related to the network: I, let us say investments, and of users' costs: This simplification is just made to enable easy explanations of the solution methods to be given. It does no harm because it will be almost always possible to use the solution methods for more realistic and complex cost functions. Quite often we will assume the benefits U to be constant, that means we will assume the total trip-matrix to be constant. Then the maximization of the social surplus changes into the minimization of the social costs.

As constraints we have the normal network constraints of Section 1.2.3 (the conservation laws, the non-negativity restrictions and the additivity). Furthermore we have the behavioural relationships of the demand model of Section 2.1.2 for the descriptive system. In the normative case these constraints do not exist. In addition we have the trip-matrix as a constraint in the case of costs minimization. Finally we may have some extra restrictions on the flows or the dimensions of the network as discussed in Section 3.2.3.

Thus we will deal with several optimization problems, stated as follows:

(a) *surplus maximization, descriptive case:*

$$\max_{\mathbf{C}} S(\mathbf{X}, \mathbf{C}) \quad \text{subject to } A\mathbf{X} = 0, \mathbf{X} \geqslant 0$$

$$G(\mathbf{X}, \mathbf{C}) = 0 \qquad (4.1.1)$$

$$H(\mathbf{X}, \mathbf{C}) >, =, < 0$$

(b) *surplus maximization, normative case:*

$$\max_{\mathbf{C}, \mathbf{X}} S(\mathbf{X}, \mathbf{C}) \quad \text{subject to } A\mathbf{X} = 0, \mathbf{X} \geqslant 0$$

$$H(\mathbf{X}, \mathbf{C}) >, =, < 0 \qquad (4.1.2)$$

(c) *costs minimization, descriptive case:*

$$\min_{\mathbf{C}} F(\mathbf{X}, \mathbf{C}) \quad \text{subject to } A\mathbf{X} = 0, \mathbf{X} \geqslant 0$$

$$G(\mathbf{X}, \mathbf{C}) = 0 \qquad (4.1.3)$$

$$H(\mathbf{X}, \mathbf{C}) >, =, < 0$$

(d) *costs minimization, normative case:*

$$\min_{\mathbf{C}, \mathbf{X}} F(\mathbf{X}, \mathbf{C}) \quad \text{subject to } A\mathbf{X} = 0, \mathbf{X} \geqslant 0$$

$$H(\mathbf{X}, \mathbf{C}) >, =, < 0 \qquad (4.1.4)$$

with:

S —total surplus

F —total costs

$\mathbf{X}$ —traffic flows (x_{ij}^{ab} or x_{ij}^{pab})

$\mathbf{C}$ —dimensions (capacities, numbers of lanes, roadwidths) of the links of the network (c_{ij})

A —matrix serving the network constraints in the case of surplus maximization (Figure 1.2.2) and serving the network constraints and the given trip-matrix in the case of costs minimization.

G(X, C)—set of functions for the description of the behaviour of the tripmakers (route choice in the case of costs minimization)

H(X, C)—set of functions for the remaining constraints.

We can write down the total social costs as:

$$F = I + T = \sum_{ij \in L} \{i_{ij}(c_{ij}) + x_{ij}t_{ij}(x_{ij}, c_{ij})\} \qquad (4.1.5)$$

We have assumed here that the 'investments' are a function of the dimension. Of course, the investments are a function of the change in the dimension (which is a function of the capital invested): $i = dc/dt$. So $i_{ij} = i_{ij}(c_{ij}^0, c_{ij})$ with $c_{ij}^0 =$ existing dimension. For reasons of simplicity we will omit this c_{ij}^0 and write $i_{ij} = i_{ij}(c_{ij})$.

Up to the present day costs minimization has received more attention (and is also easier to state and to solve) than surplus maximization. Quite often, therefore, we will discuss the costs minimization problem alone, and consider the trip-matrix as given.

4.2   MATHEMATICAL PROGRAMMING

The transport network optimization problem is a constrained optimization problem, so techniques for constrained optimization must be applied. Owing to the large number of variables and constraints, the classical approach using Lagrange multipliers cannot be applied.

The network constraints are all linear. Moreover many constraints of the set H of remaining constraints on the dimensions and/or flows will probably be linear too (for instance $c_{ij}^{min} \leqslant c_{ij} \leqslant c_{ij}^{max}$, the budget constraint

$$\sum_{ij \in L} i_{ij} \leqslant I^0,$$

in which i_{ij} is a linear function of c_{ij} and so on). So for the normative case all constraints will quite often be linear.

The probability that the objective function is linear too is not so great (owing to congestion effects for instance). However it may be possible to approximate the objective function as a piecewise linear function. In this case every decision variable which occurs in a non-linear function in the objective function, is split into a number of variables satisfying some extra constraints and adding up to the original decision variable. The original nonlinear objective function is replaced then by a linear combination of the new decision variables and a constant.

The reader is referred for an example to Figure 4.2.1, in which an original nonlinear objective function of one decision variable is replaced by a linear

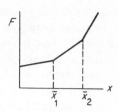

Figure 4.2.1   Piecewise linear function

objective function of three decision variables with:

$$x = \sum_{i=1}^{3} x_i$$

$$F = b + \sum_{i=1}^{3} a_i x_i$$

$$0 \leqslant x_1 \leqslant \bar{x}_1, \quad 0 \leqslant x_2 \leqslant \bar{x}_2 - \bar{x}_1; \quad 0 \leqslant x_3.$$

For convex increasing functions the solution of the linear programming problem automatically gives $x_k = 0$ unless all x_i for $i < k$ are at their upper bound. If the original objective function is not convex increasing, further measures must be taken (extra constraints imposed).

When it has linear constraints and a piecewise linear objective function, the problem has been formulated as a linear program. Theoretically this can be solved easily using the Simplex method (see for example, Hadley, 1962). For instance Carter and Stowers (1963) have stated and solved the costs minimization problem for the normative case in this way.

The big difficulty is the enormous number of variables and constraints. We remember from Section 1.2.3, that for the Dutch Integral Transportation Study the conservation laws (part of $AX = 0$) alone required 250 million constraints and 750 million variables; as all these variables may not be negative we get another 750 million constraints and so on.

Among others, Gomory and Hu (1964) and Tomlin (1966) devised methods for reducing the size of such linear programs. In this instance, use is made of the special structure of the problem and the problem is decomposed into smaller problems (see also Section 4.5.3). Moreover sometimes the problem or subproblems can be converted into problems which are easier to solve, such as the shortest path problem (see Tomlin, 1966 and also Goldman and Nemhauser, 1967). The resulting linear program is still very large for realistic networks.

Still working with linear constraints we can release the requirement of linearity for the objective function. We then get other standard optimization problems like quadratic programming and so on (see for example, Hadley, 1964). The size of the programs here is even more prohibitive.

Working with a descriptive model instead of a normative one seems preferable because a descriptive model will give a more realistic picture. The behavioural constraints $\mathbf{G}(\mathbf{X}, \mathbf{C}) = \mathbf{0}$, however, are quite complicated. Bergendahl (1969) tried to find a way out of this tricky problem by conceiving of a state of equilibrium as the solution of an optimization problem in the manner suggested in Section 2.1.3, presenting the behavioural constraints as a piecewise linear objective function and using linear programming. But he runs up against the serious problem of having two objective functions (the objective function necessary for the description of the behaviour and the original objective function for the social surplus maximization). This so-called bi-extremal problem is also not solvable by normal (linear) programming. (See Section 6.1.2 for a more complete description of the work of Bergendahl).

All methods mentioned thus for, use continuous variables. But, in most cases, the dimension of a link will not be a continuous variable but a discrete one, say 0, 2, 4 or 6 lanes; or, even more abruptly, the link is present or not present. So we get either an integer program, if all variables are restricted to being integers, or a mixed-integer program if some variables are allowed to be continuous (say $\mathbf{X}$ continuous and $\mathbf{C}$ integer).

More or less standard solution techniques exist for these problems, especially when the remaining constraints and the objective function are all linear (see for example, Hu, 1969). Roberts and Funk (1964), Hershdorfer (1965), Scott (1967) and Ventker (1970), for instance, have stated and solved the normative costs minimization problem in this way. However, once again only small networks can be managed.

It has already been said that the constraints for the descriptive cases are quite complicated. On the other hand some decision variables are not merely restricted to being integers but can only take a limited number of values (say, again, only 0, 2, 4 or 6 lanes or being present or not present). So the problem is a combinatorial one and it is possible (or necessary, if the constraints are very complex) to compute the objective function for every possible combination of these decision variables and look for the optimal solution among these combinations. When we have n links in the network and for every link m possible dimensions, the total number of combinations is m^n. This means that, for a small network with 30 links and only two possibilities for the dimension of a link (to be present or not present), there are already 2^{30}, or more than a thousand million, possible combinations. One link more means another thousand million combinations. This fact is responsible for the great difficulty of the network optimization problem. We must use very powerful techniques to tackle this difficulty. One powerful technique—though generally not powerful enough for real large networks—

is 'branch and bound', a technique for general integer programming in which the constraints can be as complicated as the traffic assignment or the whole transportation model. We will devote the whole next section to the use of branch and bound techniques in network optimization.

With respect to mathematical programming techniques we may conclude that it does not seem to be possible to solve the transport network optimization problem by a straightforward application of one of those techniques. Maybe a combination of one or more mathematical programming techniques with aggregation, decomposition (Section 4.5) and/or heuristics (Section 4.4) will be more successful.

4.3 BRANCH AND BOUND

4.3.1 The Branch and Bound Technique

'Branch and bound' is a technique in which all possible solutions of a combinatorial optimization problem are tested in a very intelligent way. Instead of computing the objective function for every possible combination of the decision variables only a small set of possible combinations have to be fully tested and the remaining possibilities can be rejected using bounding rules. Many authors give good descriptions of branch and bound, for instance Land and Doig (1960), Lawler and Wood (1966) and Mitten (1970) and of the use of branch and bound in transport network optimization, for instance Stairs (1968). Ochoa-Rosso (1968), Scott (1969b) and Scott (1970). 'Branch and bound' has been used in transport optimization by among others Ridley (1965, 1968), Perret (1967), Bureau Central d'Etudes pour les Equipements d'Outre-Mer (1967), Ochoa-Rosso and Silva (1968), Chan (1969), Scott (1969a), Pothorst (1971), Bruynooghe (1972) and Boyce and Farhi (1972).

To explain the branch and bound technique we treat the problem of minimizing $F(\mathbf{C})$ by choice of $\mathbf{C}$, where $\mathbf{C}$ belongs to the set Q of all possible feasible and infeasible solutions. To start the process we divide the set Q into n subsets Q_i, with $Q_1 \cup Q_2 \cup \cdots \cup Q_n = Q$ (branching). For every subset we compute a lower bound, that is a value F_i^l, with

$$F_i^l \leqslant F(\mathbf{C}); \qquad \mathbf{C} \in Q_i \qquad (4.3.1)$$

Furthermore we compute an upper bound F^u for the optimal solution. F^u is the value of feasible solution. We now start the bounding process by comparing F^u and F_i^l for every subset:

$$\left.\begin{array}{ll} \text{if } F_i^l > F^u & \text{the subset } Q_1 \text{ cannot include } \mathbf{C}^* \\ \text{if } F_i^l \leqslant F^u & \text{the subset } Q_1 \text{ may include } \mathbf{C}^* \end{array}\right\} \qquad (4.3.2)$$

Those subsets that cannot include the optimal solution need not to be inspected anymore. They are called inactive. The other subsets are called active and we examine them further. If we are not interested in all possible optimal solutions (with the same value for the objective function) but in just one optimal solution we may also reject the subsets with $F_1^l = F^u$, leaving only the subsets with $F_i^l < F^u$ active.

We continue the process by defining more subsets Q_{ij} by dividing one or more active subsets Q_i, again with $Q_{i1} \cup Q_{i2} \cup \cdots \cup Q_{in} = Q_i$. We again compute lower bounds F_{ij}^l and again perform the bounding process. The computation stops and the optimum solution has been found if:

$$F_i^l \geqslant F^u \quad \text{for all } i \tag{4.3.3}$$

C^* is the (or an) optimal solution with $F(C^*) = F^u$.

The efficiency of the method very much depends on the way the upper and the lower bounds are computed. Of course one must always try to have the lowest possible value for F^u and the highest possible value for F_i^l. Quite often this is accomplished by solving a simple minimization problem related to the original one.

Another important factor that influences efficiency is the branching method used. There are two important ways of branching:
(a) divide the set with the minimum lower bound;
(b) divide the set which has been reached last.
The first branching method generally requires less computing time but more computing storage, because all active subsets with their lower bound must be stored. The second alternative is less efficient in terms of computing time but more efficient in terms of storage.

In the second method a search strategy is set up in advance; we branch, as far as possible, according to the strategy, till the subsets are no longer active, then we go back as far as the first not-fully-explored subset ('node' in the 'branch and bound tree') and branch again from that node. This branching method is also called 'branch and backtrack' or 'backtrack programming' (Golomb and Baumert, 1965). The difference between the two branching methods can be shown in the 'branch and bound trees' of Figure 4.3.1. The

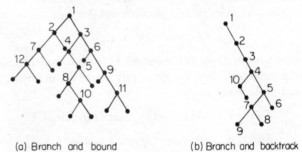

(a) Branch and bound   (b) Branch and backtrack

Figure 4.3.1   Search tree for branch and bound and
branch and backtrack

Figure 4.3.2   Flow charts for branch and bound and branch and backtrack. [Reproduced by permission of A. J. Scott and Pion Limited, from *Environment and Planning* (1969), **1**, 130 (Pion, London)]

numbers of the nodes of the trees give the order of succession of the branching. Figure 4.3.2 gives the flow charts of branch and bound and branch and backtrack. Scott (1969b) suggests a combination of the two methods. He proposes, for example, to start the process as a branch and bound process until the available storage is used up, then to switch over to backtrack programming and to return to branch and bound when there is enough storage again.

There is another very useful feature of branch and bound. Suppose one will be content with a solution that differs from the optimal one by less than 10 per cent. We can then replace the original upper bound F^u by $0.9F^u$ and can so reject more subsets Q_i (Gilmore, 1962). This, of course, reduces the amount of computing time.

To get a better understanding of the working and usefulness of branch and bound in transport network optimization problems we will discuss in the following subsections the algorithms of Ridley, Ochoa-Rosso and Silva and Chan in more detail.

4.3.2   The Method of Bounded Subsets of Ridley

Ridley (1965) was one of the first if not the first to use 'branch and bound' for transport network optimization. Only his bounding procedure is well developed, so it is not one of the most efficient methods. The problem is to minimize the total users' costs (travel time in Ridley's formulation) in a network by investing in some of the links, with a given budget as constraint. A special restriction is the fact that the costs of investing in a link are always one unit. However, it will be possible to release this restriction by slight changes in the algorithm. The linear relationship used between users' costs and investment is not in any case essential to the method. He uses a given trip-matrix and descriptive assignment according to the shortest-route principle.

So the problem is stated as follows:

$$\min_{y_{ij}} F = \sum_{ij \in L} x_{ij} t_{ij} \qquad (4.3.4)$$

subject to:

the network constraints (1.2.11), (1.2.12) and (1.2.13)
a fixed trip-matrix (2.2.21)
the shortest-route principle (2.1.5)

$$t_{ij} = a_{ij} - b_{ij}y_{ij} \quad \text{for all } ij \in L$$
$$y_{ij} = 0 \quad \text{for } ij \in \bar{L}_I \quad \text{and} \quad y_{ij} = 0 \text{ or } 1 \quad \text{for } ij \in L_I$$

$$\sum_{ij \in L} y_{ij} i_{ij} \leqslant I^0$$

$$i_{ij} = 1 \quad \text{for } ij \in L$$

with:

y_{ij}—Boolean variable, which indicates if an investment is made in link ij:
$y_{ij} = 0$ means no investment made in link ij
$y_{ij} = 1$ means an investment made in link ij
L_I—set of links, which may be invested in
$\bar{L}_I$—set of links, which may not be invested in
I^0—fixed budget for investments

We start to minimize F ignoring the budget constraint. The total investments ($=$ number of investments, because $i_{ij} = 1$) are then say I^{max}.

$$I^{\text{max}} = \sum_{ij \in L_I} i_{ij}$$

Next we minimize F with a budget constraint of $I^{\text{max}} - 1$, so it cannot be invested in one road. This procedure is continued until the real budget constraint of I^0 is reached:

$$F_{I^{\text{max}}} \leqslant F_{I^{\text{max}} - 1} \leqslant \cdots F_m \cdots \leqslant F_{I^0 + 1} \leqslant F_{I^0}$$

Ridley gives some bounding rules to compute F_m in a simple way, having done the computation for $m + 1$. The set Q of all solutions with m investments is divided into the $\binom{n_{L_I}}{m}$ subsets of all possible combinations for the m investments (with n_{L_I}—numbers of links, which may be invested in). Let us define the network A_m as such a network and A_{m+1} as a network with one investment more. Then, of course, the following holds:

$$F(A_m) \geqslant F(A_{m+1})$$

We now state a lower bound for $F(A_m)$:

$$F^l(A_m) = \max_{A_{m+1} \supset A_m} F(A_{m+1})$$

(The notation $A_{m+1} \supset A_m$ means 'A_m is contained in A_{m+1}'). In addition we have to compute an upper bound F_m^u for F_m. For this upper bound we always use the lowest value of F_m computed. As a starting value $F_m^{u(0)}$ for F_m^u we use:

$$F_m^{u(0)} = F(A_m^{(0)})$$

in which $A_m^{(0)}$ is a network defined as follows:

$$F^l(A_m^{(0)}) = \min_{A_m} F^l(A_m)$$

It seems plausible that this network will also give a low value for the upper bound. The procedure to obtain F_m using the results of the computations of F_{m+1} is shown in the inner part of the flow chart of Figure 4.3.3. So instead

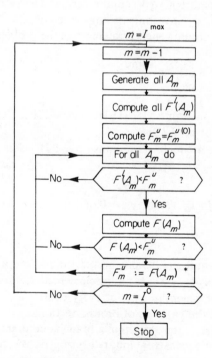

Figure 4.3.3   Flowchart of the method of bounded subsets of Ridley
[Reproduced by permission of Microforms International Marketing
Corporation, from *Transportation Research* (1968), **2**, 420]
* The notation := has been borrowed from **ALGOL** and has the
meaning 'to be replaced by'

of computing $F(A_m)$ for all combinations (which should mean a traffic assignment for each combination) only a few networks must be explored fully; most networks can be rejected by the bounding rule.

Note that for the computation of $F^l(A_m)$ it is necessary to know all $F(A_{m+1})$. This would mean that we would still have to do all the assignments, only one step further on. To avoid that we use $F^l(A_{m+1})$ instead of $F(A_{m+1})$. The price we have to pay is that the lower bounds become lower and the bounding rule less strong. Therefore we will need to compute $F(A_m)$ more often.

In computing the optimal network for a budget I^0 we also get all the intermediate optimal networks for the budgets $I^0 + 1$ until I^{max}. This, of course, is a very interesting by-product.

4.3.3   The Branch and Bound Algorithm of Ochoa-Rosso and Silva

Ochoa-Rosso and Silva (1968) developed a 'real' branch and bound algorithm for transport network optimization. Their objective was users' costs minimization by investing in links, subject to a budget constraint. They give a slightly more general statement of the problem than Ridley did.

$$\min_{y_{ij}} F = \sum_{ij \in L} x_{ij} t_{ij} \qquad (4.3.5)$$

Subject to:

the network constraints (1.2.11), (1.2.12) and (1.2.13)
a fixed trip-matrix (2.2.21)
a route choice model [e.g. relationship (2.1.5)]⎫ F is monotonically de-
                                                ⎬   creasing when projects
$t_{ij} = t_{ij}(y_{ij})$ ⎭   are being implemented

$y_{ij} = 0$ for $ij \in \bar{L}_I$ and $y_{ij} = 0$ or 1 for $ij \in L_I$

$\sum_{ij \in L} x_{ij} i_{ij} \leqslant I^0$

Although the requirement that the total users' costs are monotonically decreasing when projects are being implemented seems very plausible and normal, this is not generally true for every route choice model and users' costs function (see the paradox of Braess, Section 2.2.2.2).

During the course of the algorithm a branch and bound tree is generated. The nodes of this tree consist of mutually exclusive but collectively exhausting subsets Q_i of the set Q of all solutions. The branching operation consists of the division of a subset Q_i into two subsets Q_{i1} and Q_{i2}. At the start all variables y_{ij} with $ij \in L_I$ are free to be chosen 0 or 1 (the set Q). At every branching operation one (more) variable y_{ij} is fixed to $0 (Q_{i1})$ or $1 (Q_{i2})$.

For every subset Q_i a lower bound F_i^l for the solution contained in Q_i can be computed by setting all free variables equal to 1 (due to the requirement that F is monotonically decreasing). We can also inspect the subset to see if it contains any feasible solutions by setting all free variables equal to 0. If a subset does not contain any feasible solution we do not need to consider it any longer and the node becomes inactive. Moreover it may be possible to set an upper bound F^u on the optimal solution. Though this is not absolutely necessary for the algorithm of Ochoa-Rosso and Silva, it can increase the efficiency considerably. It can be done, for instance, by computing F with projects ranked in order of their investment costs and accepting as many as possible without exceeding the budget. If $F_i^l \geqslant F^u$ we need not consider Q_i any longer and again Q_i is no longer active. We branch now from that active node with the lowest value for F_i^l. Before doing so we check to see if the solution which gives the lowest value for F_i^l is feasible. If so, the optimal solution has been obtained. In Figure 4.3.4 a flow chart is given for the elementary branch and bound operation of Ochoa-Rosso and Silva.

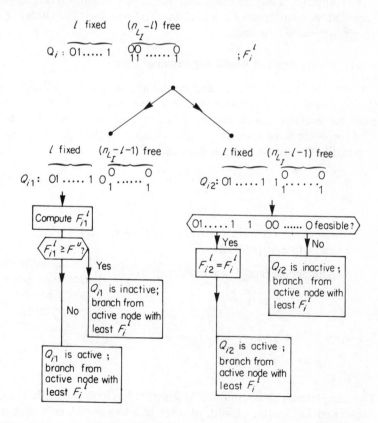

Figure 4.3.4   Flow chart of the elementary branch and bound operation of Ochoa-Rosso and Silva

In node i the set Q_i consists of l fixed y_{ij}'s and $n_{L_I} - l$ free variables. In the branching operation one more variable is fixed and the two resulting subsets S_{i1} and S_{i2} are inspected. The nodes $i1$ and $i2$ can be active or inactive, because the sets Q_{i1} and Q_{i2} may or may not contain feasible solutions with a value for the objective function lower than the upper bound.

It is obvious that, for the calculations of a F_i^l, a traffic assignment is needed. As is known (and will also be shown in Chapter 7) this takes a lot of computing time. Ochoa-Rosso and Silva remark that the network of node $i1$ differs only in one link from the network of node i. In such a case it is possible to use fast procedures to compute the total users' costs (at any rate when an all-or-nothing assignment according to the shortest route is used) (see further Section 7.7). Moreover it is not necessary to compute the values F_i^l at the beginning of the process. For the node with the most 1's for the y_{ij}'s will always have the least value for F_i^l. So we continue branching from the node with most projects implemented until the budget constraint is reached. Only from that moment onward is it necessary to know the F_i^l's in order to decide from which node to branch.

4.3.4   The Branch and Bound Algorithm of Chan

Chan (1969) uses a branch and bound algorithm, very similar to that of Ochoa-Rosso and Silva, to find the least investment costs solution to satisfy a certain level of effectiveness, measured in total users' costs. The statement of the problem is the same as in Section 4.3.3, only the role of objective function and budget constraint has been changed;

$$\min_{y_{ij}} F = \sum_{ij \in L} y_{ij} i_{ij} \qquad (4.3.6)$$

subject to:

the network constraints (1.2.11), (1.2.12) and (1.2.13)
a fixed trip-matrix (2.2.21)

a route choice model [e.g. relationship (2.1.5)] $\left.\begin{array}{l} \\ \\ \end{array}\right\}$ T is monotonically decreasing when projects are being

$t_{ij} = t_{ij}(y_{ij})$ $\left.\begin{array}{l} \\ \end{array}\right\}$ implemented

$y_{ij} = 0$ for $ij \in \bar{L}_I$ and $y_{ij} = 0$ or 1 for $ij \in L_I$

$$\sum_{ab \in P} x^{ab} t^{ab} \leqslant T^0$$

The same branch and bound tree is generated as in the algorithm of Ochoa-Rosso and Silva, starting with all variables free and then branching each time by fixing one more variable. For every subset Q_i a lower bound F_i^l for the total investments is found by setting all free variables equal to zero.

Now Q_i either may, or may not, contain feasible solutions. To check this, we proceed to compute the total users' costs T_i for a network in which all free variables are set equal to one. If no feasible solutions are forthcoming, subset Q_i need not be considered further. When an upper bound for the solution F^u is available, we can also reject all subsets for which $F_i^l \geqslant F^u$.

We now branch from the active node with the least value for F_i^l. For the termination Chan proposes to continue the computation till all subsets are inactive or consist of only one feasible solution. The optimal solution is then the feasible solution with the least value for the objective function. For reasons of efficiency a strong low upper bound F^u is then required.

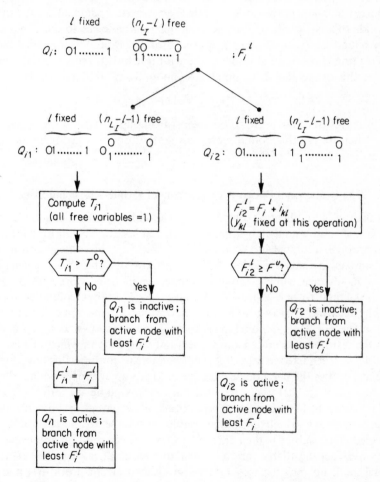

Figure 4.3.5   Flow chart of the elementary branch and bound operation of Chan

Another termination rule could be a similar rule to that used by Ochoa-Rosso and Silva : always branch from the active node with the least value for F_i^l and always check if F_i^l is feasible. The optimal solution has been obtained the first time F_i^l is feasible. However, this would mean a computation each time of the total users' costs with all free variables set equal to zero (in the right branch of Figure 4.3.5). As in the method of Ochoa-Rosso and Silva we need to compute the total users' costs quite a few times. The same remark can be made here as in the preceding section, namely that the network for T_{i1} differs only in one link from the network for T_i. So it is fairly easy to compute T_{i1} having computed T_i, when an all-or-nothing assignment according to the shortest route is used (see Section 7.7).

Moreover Chan proposes a so-called 'arithmetic update'. He says that to decide that Q_{i1} contains feasible solutions it is sufficient to show that an upper bound for $T_{i1} \leqslant T^0$. An upper bound for T_{i1} is found by assuming that no rerouting takes place after fixing the y for one link at zero. So, assuming that in the branching from node i to node $i1$, y_{kl} is fixed, we obtain :

$$T_{i1} \leqslant T_i - x_{kl}^{(i)}\{t_{kl}(y_{kl} = 1) - t_{kl}(y_{kl} = 0)\}$$

However this does not seem to be a very strong upper bound for T_{i1} especially when we use this bound repeatedly so that only the upper bound for T_i is available instead of T_i itself.

4.3.5   The Branch and Backtrack Algorithm of Ochoa-Rosso and Silva

In their 1968 publication, Ochoa–Rosso and Silva present, in addition to their branch and bound algorithm, a branch and backtrack algorithm for the same problem as stated in Section 4.3.3 (Problem (4.3.5)). This algorithm is very similar to that of Section 4.3.3. Only now the branching scheme is defined in advance, while F_i^l is only computed when the network with all free variables equal to one, is feasible. Because this F_i^l is then computed on a feasible network, it may also yield a new value for the upper bound F^u (if $F_i^l < F^u$). The general branching scheme is now as follows : always branch from the last node generated and inspect only the node in which one more variable has been fixed to zero. If it is not possible to continue this branching scheme, because the last node is inactive or yields a new upper bound, then it is necessary to backtrack. Backtracking means going up into the tree, until reaching the first node that has not been fully explored, i.e. for which the succeeding node in which one more variable has been set to one, has not been inspected. That subset is then checked to see if it can contain any feasible solutions by setting all free variables equal to zero. If not, it is again necessary to backtrack up into the tree. Otherwise, branch further from that node, according to the general branching scheme. It is not now possible to terminate the search before all subsets have been totally inspected. Figure 4.3.6 gives

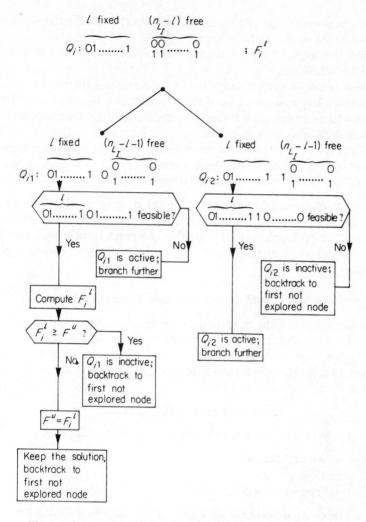

Figure 4.3.6 Flow chart of the elementary branch and backtrack operation of Ochoa-Rosso and Silva

a flow chart for the elementary branch and backtrack operation of Ochoa-Rosso and Silva.

The main differences exhibited by the branch and backtrack algorithm when compared with the branch and bound algorithm are:

(a) the branching scheme has been fixed in advance and so the branching is not the most efficient;

(b)   one can only be sure that the optimal solution has been found when every subset has been totally explored;

(c)   only the last solution and the feasible solution with least upper bound need to be stored;

(d)   there is always (after a short time) a good estimation for F^u;

(e)   perhaps less computations of F_i^l are needed; on the one hand F_i^l need only to be computed if it is also a feasible solution, on the other more subsets S_i need to be inspected;

(f)   when for some reason, the computation must be stopped before the optimum has been found, there is always (after a short time) a feasible solution, which, it is reasonable to hope, is a fairly good one.

Ochoa-Rosso and Silva state that the choice between the two types of algorithm must be made according to the kind of network and other properties of the problem, and the available computer storage and computer time. Only experiment will enable one to select the appropriate type of algorithm.

4.3.6   A Branch and Bound Algorithm for the General Problem

The algorithms of Ridley, Ochoa-Rosso and Silva and Chan already discussed are all methods for the minimization of one part of the total costs (users' costs or investment costs) for a fixed trip-matrix. It is not difficult however to solve the general problem of surplus maximization in the same way. Let us first state the problem:

$$\max_{\mathbf{Y}} \{U(\mathbf{Y}) - T(\mathbf{Y}) - I(\mathbf{Y})\} \qquad (4.3.7)$$

subject to the usual constraints and a transportation model, with:

  U—total users' benefits
  T—total users' costs
  I—total investment costs.

As is known from Section 3.2.2.2 it is only possible to give an estimate of the difference of $S = U - T$ of one network with respect to another one. Let us define the network with all $y_{ij} = 0$ as network of reference. When the problem is not a problem of network improving but of defining a totally new network, a good network of reference would be the network that connects all origins and destinations and that has the least investment costs (the problem of the minimal spanning tree). The investment costs for that network also serve as a good general lower bound for the investment costs of all networks.

It is reasonable to suppose, then, that ΔU is a positive and monotonically increasing function and ΔT is a negative and monotonically decreasing function, so ΔS is a positive and monotonically increasing function when

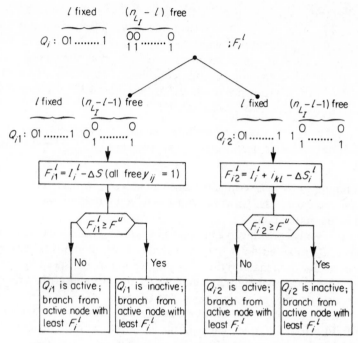

Figure 4.3.7 Flow chart of the elementary branch and bound operation
for the general problem

projects are being implemented. Let us now reformulate the problem:

$$\min_{y_{ij}} F = \sum_{ij \in L} y_{ij} l_{ij} - \Delta S(y_{ij}) \tag{4.3.8}$$

subject to:

the network constraints (1.2.11), (1.2.12) and (1.2.13)

a transportation model $\left.\begin{array}{l} \\ \\ t_{ij} = t_{ij}(y_{ij}) \end{array}\right\}$ ΔS is monotonically increasing when projects are being implemented

$y_{ij} = 0$ for $ij \in \bar{L}_I$ and $y_{ij} = 0$ or 1 for $ij \in \bar{L}_I$

We build now the same branch and bound tree as Ochoa-Rosso and Silva and Chan. So we start with all variables y_{ij} free and fix one variable more when we branch from a node. A lower bound F_i^l for each subset is easily found by:

$$F_i^l = \sum_{\substack{\text{all fixed } y_{ij} \text{ have} \\ \text{their fixed values} \\ \text{and all free } y_{ij} = 0}} y_{ij} l_{ij} - \Delta S \quad \text{(all fixed } y_{ij} \text{ have their fixed values and all free } y_{ij} = 1)$$

As well as the lower bounds it is important to have a very good (i.e. low) upper bound F^u. This upper bound is the value of the objective function for a feasible solution. It is sensible to choose for this feasible solution at the start of the process a network which seems to be a very good one and to try to improve this F^u when the computation continues. The elementary branch and bound operation is shown in the flow chart of Figure 4.3.7. For simplification we use in the figure two new notations:

I_i^l—total investment costs with all free variables of node i equal to 0.

ΔS_i^l—total change in users' surplus with all free variables of node i equal to 1.

The computation can be terminated when all subsets are inactive or consist of only one network. The network with the lowest value for the objective function is the optimal one.

Of course it is also possible to construct a branch and backtrack algorithm for the general problem similar to that of Ochoa-Rosso and Silva. The same remarks that have been made in the preceding sections about speeding up the computation by using special reassignment techniques and arithmetic updating to compute the total users' costs, can be made here.

The branch and bound algorithm presented above has not been tested on a network. But from the resemblance of the algorithm with those of Ochoa-Rosso and Silva and Chan one may suppose that it would be as efficient as theirs, at least when a good upper bound is available.

4.3.7   Discussion of Branch and Bound Techniques

As shown in the preceding sections it is possible with the help of branch and bound to solve the general transport network optimization problem, and such related problems as users' costs minimization, for almost all possible kinds of functional relationships between the variables. Even a complex set of functions such as the transportation model can be handled. The only restriction is that it must be possible to set lower bounds for the solution contained in each subset of solutions and an upper bound for the optimal solution. For this reason a monotonic relationship between the parts of the objective function and the decision variables is very handy but not absolutely necessary.

The optimal solution is always found after a finite number of steps (at least when the subsets are mutually exclusive: $Q_i \cup Q_j = \phi$). Furthermore we quite often know at each step of the process a feasible solution and a lower and upper bound for the optimal solution. Finally one may generally say that branch and bound is a clear and mathematically elegant way of solving the network optimization problem.

The most important point now, of course, is the efficiency of the method. We note with regret that it is always necessary to compute the total users costs quite a few times. To do this it is generally necessary to compute an assignment to the network. As shown in Section 2.1.4.2 and as will be shown further in Chapter 7 this takes a lot of computing time even for moderately sized networks, especially when a capacity restraint technique is used. When we use an all-or-nothing assignment according to the shortest route it is possible to simplify the computation of the users' costs considerably. It will still be quite often necessary to do an assignment.

This feature of the branch and bound methods probably makes it prohibitive to use them for larger networks (see also Stairs, 1968). As far as we know branch and bound has until now only been used in practice directly for networks of up to twenty or thirty nodes. However, in spite of the fact that branch and bound is not directly of use for larger networks, it might be very useful in combination with other techniques like heuristic techniques (Section 4.4) or decomposition and aggregation (Section 4.5; see also Chan, 1969). Also we can adopt some of the concepts of branch and bound without adopting the whole method.

4.4 HEURISTIC TECHNIQUES

4.4.1 The Nature of a Heuristic Program

The problem of transport network optimization for realistic networks has two important general features:

(a) *the size of the problem is very large*; this means that there are many variables and many constraints;
(b) *it is a very complex problem*; the relationships between the variables (the constraints) may be as complex as the whole transportation model but at any rate as complex as a descriptive assignment.

These two features have some important consequences. One of them is that it may be impossible to find an optimal solution within a reasonable computing time. The objective may then change from finding an optimal solution to finding a feasible solution, that is 'fairly' good or 'near optimal'. Sometimes we must be even content with a solution that is merely feasible. When we do not look for the optimal solution it is generally very nice to know something of it, say an upper or a lower bound. As shown in the treatment of branch and bound techniques we can sometimes find such a bound by solving a related but easier problem; for instance, by ignoring integrity constraints or by assuming relationships linear which are not or by neglecting interdependencies and so on. When we know such a bound, we can also estimate how 'near optimal' the feasible solution we have found is.

Looking for solving methods for such large and complex problems, in which only a fairly good solution is needed, we examine the problem-solving process in human brains (see, for instance, Bergendahl, 1971). We can then see some features, that could be 'built in' in a computer solving process, for example

> simplifying the problem by ignoring some constraints or by replacing complex constraints with simpler ones;
> splitting the complex main problem into easier subproblems;
> neglecting the interdependency of subproblems;
> finding an initial feasible solution:
> trying to improve this initial solution in an iterative process (so-called hill-climbing or -descending procedures);
> judging whether the improvement is sufficiently good;
> deciding to stop or to continue the searching process;
> rejecting immediately solutions or sets of solutions that are not promising.

By putting these problem-solving procedures into a computer program that carries out the solution process, we get advantages such as:

> the simplifying, selecting and deciding rules are explicitly and uniquely defined;
> the computer executes simple calculations much faster than the human brain;
> the storage of a computer is bigger than that of the human brain;
> the experiences in problem solving of many people can be put into one computer program.

In such a way we can try to formulate so called 'heuristic' programs which can give reasonable solutions for such large and complex problems as the network optimization problem for realistic cases within a reasonable computing time. In the next sections we will discuss some heuristic programs.

Although a heuristic program is quite often the only available solution method, it is also a very dangerous one; so we must be very careful in using it. Quite often one does not know how far the solution found is from the optimal one. Even if we have a good estimate for the distance between the solution found and the optimal solution for the objective function, we do not have one for the decision variables. This distance can be large, even if the value of the objective function is not too far from the optimal value.

Moreover some selecting and deciding rules seem very plausible and good when, in reality, they are not. Remember, for instance, the paradox of Braess (Section 2.2.2.2) in which it is shown that an increase in the number of links can also mean an increase in the total users' costs for a fixed trip-matrix. So it is quite possible that, with a heuristic method, we can be going in the wrong direction thinking it to be the right one. We will give some examples of this phenomenon in the following sections.

4.4.2 Neglecting the Interdependency in the First Instance

4.4.2.1 Continuous Optimal Adjustment

We will deal with the problem of costs minimization for a given trip-matrix and with a descriptive assignment:

$$\min_{C} F(X, C) \tag{4.4.1}$$

subject to:

$$AX = 0, \quad X \geqslant 0$$

$$G(X, C) = 0$$

or else stated:

$$\min_{c_{ij}} \sum_{ij \in L} F_{ij}(x_{ij}, c_{ij}) \tag{4.4.2}$$

subject to: the network constraints and a fixed trip-matrix
          a route choice model.

Steenbrink (1970a, 1971) proposed a method of solution by ignoring in turn the constraints and the optimization. So we start with an assignment of the trip-matrix to a starting network, that means we start with the constraints. Then we get a traffic flow for each link: $x_{ij}^{(1)}$. Next we solve the constraintless problem:

$$\min_{c_{ij}} \sum_{ij \in L} F_{ij}(x_{ij}^{(1)}, c_{ij}) \tag{4.4.3}$$

This is nothing more than n_L times the mathematically simple problem:

$$\min_{c_{ij}} F_{ij}(x_{ij}^{(1)}, c_{ij}) \tag{4.4.4}$$

because F_{ij} is independent of c_{kl} for $ij \neq kl$. This last problem—the optimal dimension for a given traffic flow—can be solved by inspecting all possibilities, by a search method or by differentiating when we assume $F_{ij}(c_{ij})$ to be a differentiable function of the continuous variable c_{ij} (see also Section 10.1).

Note that in the solution of problem (4.4.3) the network constraints and the trip-matrix constraints are still met, but not the route choice model. The resulting network $\{c_{ij}^{(1)}\}$ is now used as a new network to which the trip-matrix is assigned. So the problem is now:

$$\text{find an } X^{(2)}, \text{ such that } AX^{(2)} = 0, X^{(2)} \geqslant 0$$
$$G(X^{(2)}, C^{(1)}) = 0. \tag{4.4.5}$$

After the assignment a constraintless optimization is again executed:

$$\min_{c_{ij}} \sum_{ij \in L} F_{ij}(x_{ij}^{(2)}, c_{ij}) \qquad (4.4.6)$$

This process is repeated till the changes in the objective function are too small to justify continuation of the process.

Ventker (1970) proposed a similar procedure. It is different only in that the process is started by a costs minimization with a normative assignment. Next the trip-matrix is assigned to this perhaps fairly good starting network and the same procedure as given above is executed.

Ignoring the effect of Braess this process is indeed a hill-descending procedure. For:

$$\sum_{ij \in L} F_{ij}(x_{ij}^{(n)} c_{ij}^{(n)}) \leqslant \sum_{ij \in L} F_{ij}(x_{ij}^{(n)}, c_{ij}^{(n-1)})$$

due to the minimization procedure. And:

$$\sum_{ij \in L} F_{ij}(x_{ij}^{(n)}, c_{ij}^{(n-1)}) \leqslant \sum_{ij \in L} F_{ij}(x_{ij}^{(n-1)}, c_{ij}^{(n-1)})$$

due to the effect of minimizing users' costs in a descriptive assignment (though there is a difference between the total users' costs in a normative and a descriptive assignment, we may generally suppose that the total users' costs decrease when a proper descriptive assignment is made).

So this process seems to lead to fairly good solutions in which every step of the process does at least give an improvement. It is also an advantage that the process is very easy to understand. Moreover we see that this process takes place in reality, not as an iterative computing scheme but in the form of real decisions and real behavioural responses to these decisions. New investments are made in a road. Therefore this road becomes more attractive for the users, so the traffic flow on that road increases. Due to that fact further investments are needed and the process continues.

This process is a typical example of a heuristic procedure:

> it is a simulation of a certain human decision process; it gives an improvement in every step of the computation, so it seems to lead to a fairly good solution;
> it is not guaranteed that the optimum is found.

It is very illustrative to show that this heuristic process can lead to quite wrong solutions. As an example let us take a network consisting of two roads (Figure 4.4.1). Road 1 is short in distance but expensive in expanding costs; road 2 has a long distance but is cheap to expand. Let us suppose that in the starting solution both roads are present. Road 1 then gets more traffic than road 2 due to the lower users' costs. So it becomes necessary to expand road 1, though it is expensive. After this expansion, in the next step of the computing

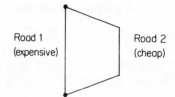

Road 1
(expensive)

Road 2
(cheap)

Figure 4.4.1 Network for the example of the
method of continuous optimal adjustment

process road 1 has become even more attractive for the users, so the traffic flow on road 1 increases. Due to that effect it becomes again necessary to invest in road 1, though it is expensive. The solution of expanding road 2 (possibly cheaper socially), which would increase users' costs but reduce investment costs is never found. If, in another starting solution, road 1 or road 2 were to possess so few lanes that the road carried no traffic in the starting solution, the road would never get any traffic (if the other road is not narrowed).

We can put the phenomenon described above in another way. Say an investment in a road has two effects. The primary effect is the improvement of the situation for the traffic flow already on the road. The secondary effect is a change in the route choice. In the method of the continuous optimal adjustment these secondary effects are ignored in the first instance. This is only justifiable when the secondary effects are significantly smaller than the primary effects, which is generally not true of traffic flows on a transport network.

In Section 6.2.3 this method is applied to a two-mode transport network optimization problem. There too the phenomenon described above occurs.

4.4.2.2 Selecting the Most Promising Projects

Another heuristic technique is to select the most promising projects and to invest in them. This technique too is very similar to human decision-making. We will deal again with the problem of costs minimization for a given trip-matrix and with a descriptive assignment:

$$\min_{c_{ij}} F = \sum_{ij \in L} \{i_{ij}(i_{ij}) + x_{ij}t_{ij}(x_{ij}, c_{ij})\} \qquad (4.4.7)$$

subject to:

the network constraints and a fixed trip-matrix
a route choice model.

As in Section 4.4.2.1, we ignore in turn the optimization and the route choice model constraint. Starting by ignoring the optimization we get the descriptive assignment $\mathbf{X}^{(1)}$ to a starting network. Ignoring the constraints we now get the minimization problem:

$$\min_{c_{ij}} F^{(1)} = \sum_{ij \in L} \{i_{ij}(c_{ij}) + x_{ij}^{(1)} t_{ij}(x_{ij}^{(1)}, c_{ij})\} \qquad (4.4.8)$$

Assuming for a moment that $F^{(1)}$ is differentiable with respect to the continuous variables c_{ij}, in the minimum (4.4.9) holds:

$$\frac{\partial F^{(1)}}{\partial c_{ij}} = 0; \quad \text{for all } ij \in L; \qquad (4.4.9)$$

or

$$\frac{\mathrm{d}F_{ij}^{(1)}}{\mathrm{d}c_{ij}} = 0; \quad \text{for all } ij \in L; \qquad (4.4.10)$$

because $F_{ij}^{(1)}$ is independent of c_{kl} for $ij \neq kl$. Because of the convexity of $F_{ij}^{(1)}$ (see Chapter 2) we know that:

$$\frac{\mathrm{d}F_{ij}^{(1)}}{\mathrm{d}c_{ij}} < 0 \quad \text{if } c_{ij} < c_{ij}^* \qquad (4.4.11)$$

so we have to raise the dimension of link ij; this means investing in link ij if $\mathrm{d}F_{ij}^{(1)}/\mathrm{d}c_{ij} < 0$ and continuing this process till the differential-quotient equals zero. This means investing if:

$$\frac{\mathrm{d}i_{ij}}{\mathrm{d}c_{ij}} + x_{ij}^{(1)} \frac{\mathrm{d}t_{ii}}{\mathrm{d}c_{ij}} < 0 \qquad (4.4.12)$$

or:

$$\frac{-x_{ij}^{(1)} \dfrac{\mathrm{d}t_{ij}}{\mathrm{d}c_{ij}}}{\dfrac{\mathrm{d}i_{ij}}{\mathrm{d}c_{ij}}} > 1 \qquad (4.4.13)$$

This is the well-known rule of investing in a project if the reduction of users' costs outweighs the investment costs. We may call the reduction of the users' costs the benefit of the investment and the investment costs the costs of the investment. So we get the rule to invest if the value of the benefits minus the costs is positive or if the benefit/cost ratio is greater than one. This rule also holds when $F_{ij}^{(1)}$ is not differentiable and c_{ij} is not continuous. Using this rule directly we have the same method as described in the preceding section. But it is also possible not to invest in all projects with $-x(\mathrm{d}t/\mathrm{d}c)/(\mathrm{d}i/\mathrm{d}c) > 1$, but only in the ones with the highest value for the benefit/cost ratio and only to continue the investments up to a point where the ratio is still higher than one. This is especially relevant when a budget constraint is used. Then the rule is: invest in those projects with highest benefit/cost ratio till the budget constraint is reached. This last method has been used for instance by the Bureau Central d' Etudes pour les Equipments d' Outre-Mer (Villé, 1969).

It is clear that the same kind of danger is present here as in the method of the continuous optimal adjustment. So, among others, Pearman (1971) warns against the use of methods like this. We will show in an example similar to one given by Pearman how wrong the results obtained by this method can be. We will use the network of Figure 4.4.2, and a trip-matrix consisting of two relations; 13 and 14.

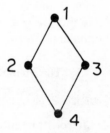

Figure 4.4.2   Network for the example of the method of selecting the most promising projects

We assume $x^{13} = 500$ and $x^{14} = 2,000$, assume the users' costs consisting of evaluated travel time costs (with a value of travel time of 1/3·5) and use the shortest route principle for the assignment. In Table 4.4.1 we give the users' costs on the links before and after an investment is made, the investment costs, the $x_{ij}^{(1)}$ and the benefit/cost ratio per link. So it is very easy to

Table 4.4.1   Example of the method of selecting the most promising projects

Link	t^{before}	t^{after}	i	$x^{(1)}$	benefit/cost ratio
12	20/3·5	15/3·5	1,000	2,000	10/3·5 = 2·857
24	20/3·5	18/3·5	1,000	2,000	4/3·5 = 1·143
13	21/3·5	15/3·5	1,000	500	3/3·5 = 0·857
34	20/3·5	17/3·5	1,000	0	0/3·5 = 0

decide what to do, namely to invest in the links 12 and 24. Let us compute the objective function before and after the investments have been made:

$$F \text{ (no investment)} \qquad = 90,500/3·5$$

$$F \text{ (invested in 12 and 24)} = 83,500/3·5$$

So we have indeed got an improvement. When we make an assignment to the improved network we get the same distribution of traffic flows and no further improvement can be made. However when we invest in the projects with a value of the benefit/cost ratio smaller than one, namely the links 13 and 34, we get a better value for the objective function:

$$F \text{ (invested in 13 and 34)} = 78,500/3 \cdot 5$$

This lower value of the objective function is caused by the change in the assignment due to the change in users' costs.

From the examples of the Sections 4.4.2.1 and 4.4.2.2 we may draw the conclusion that it can be very dangerous to use simple and plausible heuristic methods to select the best investment projects. Though we get an improvement at every step of the computation, the optimum is only occasionally reached and we may well go in a completely wrong direction. So methods like this are of hardly any use, except to make an adjustment in a nearly optimal network.

4.4.3   The Method of Barbier

4.4.3.1   The Original Method of Barbier

Barbier presented, in 1966, a heuristic method for the costs minimization problem with a descriptive assignment (according to the shortest route). For realistic networks this method gives a fairly good solution in a reasonable computing time. Barbier has used his method for a study of possible extensions to the network of the Metro in Paris, in which he used 36 origins and destinations and a network consisting of about 60 nodes and about 280 links. Haubrich (1972, Nederlands Economisch Instituut 1972, annex VI) has used an extended version of the method of Barbier for the optimization of the railway networks in the Dutch Integral Transportation Study. This network consisted of about 1,250 nodes and about 8,000 links and about 450 origins and destinations were used. On an IBM 360/65 computer the solution was found within 40 minutes.

The method of Barbier is a switching process like the method of the continuous optimal adjustment. First a descriptive assignment is made to a starting network with all possible links present. Then an inspection is made for every link ij to see if it is possible to assign the traffic flow originally assigned to that link ij to other links in such a way that a lower value for the objective function is obtained. If so the link ij originally used is ejected from the network. By the time this inspection has been completed for all links, a new network has been obtained. A descriptive assignment is made to this new network and the process is repeated, We will give a more detailed description of the method of Barbier below.

The problem can be stated as follows:

$$\min_{y_{ij}} \sum_{ij \in L} \{ y_{ij} i_{ij} + x_{ij} t_{ij}(y_{ij}) \} \tag{4.4.14}$$

subject to:

$AX = 0, X \geqslant 0$ (the network constraints and a fixed trip-matrix)

$G(X, Y) = 0$ (a route choice model with $x_{ij} = 0$ if $t_{ij} = \infty$)

$y_{ij} = 0$ for $ij \in \bar{L}_I$ and 0 or 1 for $ij \in L_I$*

$i_{ij} = 0$ for $ij \in \bar{L}_I$

$t_{ij}(1) \leqslant t_{ij}(0)$ for all $ij \in L_I$

The solution method is an iteration process; we give the steps of the process below:

Step 1. Assign the trip-matrix to the maximum possible network, so a network in which $y_{ij} = 1$ for all $ij \in L_I$, according to the descriptive route choice principle. The result of this assignment is a traffic flow $x_{ij}^{(1)}$ on each link, such that all constraints are satisfied.

Step 2. In the second step we try to minimize the objective function by 'rerouting' the traffic flows originally assigned to the links. We might say that we consider a new trip-matrix $x^{ij} = x_{ij}^{(1)}$, and that we assign that matrix to the network according to the route with lowest value for the objective function. We then get a second vector of traffic flows $X^{(2)}$ and a vector $Y^{(2)}$ such that $y_{ij}^{(2)} = 0$ if $x_{ij}^{(2)} = 0$ and $y_{ij}^{(2)} = 1$ if $x_{ij}^{(2)} > 0$. The assignment is made in the following way:
(a)   put $x_{ij}^{(2)} = 0$ and $y_{ij}^{(2)} = 0$ for all $ij \in L$; and $t_{ij} = t_{ij}(1)$ for all $ij \in L_I$
(b)   then, for all $ij \in L$:
   find a path from i to j such that:

$$\sum_{\substack{kl \in p \\ p \text{ forms a path from } i \text{ to } j}} \{i_{kl}(1 - y_{kl}^{(2)}) + t_{kl}x_{ij}^{(1)}\} \qquad (4.4.15)$$

is minimized and, for all links kl contained in the minimum path from i to j:

$$x_{kl}^{(2)} := x_{kl}^{(2)} + x_{ij}^{(1)};$$
$$y_{kl}^{(2)} := 1. \qquad (4.4.16)$$

When the descriptive assignment of the preceding step has been made according to the route with minimum users' costs, we can simplify the above reassignment process because we know that $i_{ij} = 0$ for $ij \in \bar{L}_I$ and so ij will be the path from i to j with the minimum value for the objective function, consisting of investments and users' costs.

---

* If more investment levels for a connection ij must be considered, we need to construct one 'link' between i and j for each investment level.

The assignment then becomes:

(a) put $x_{ij}^{(2)} = x_{ij}^{(1)}$ for all $ij \in \bar{L}_I$

$x_{ij}^{(2)} = 0$ for all $ij \in L_I$

$y_{ij}^{(2)} = 0$ for all $ij \in L_I$

$t_{ij} = t_{ij}(1)$ for all $ij \in L_I$ (4.4.17)

(b) then, for all $ij \in L_I$:

perform the same path finding and assignment procedure as defined above in relations (4.4.15) and (4.4.16).

When the reassignment process has been completed we have a new network, described by the vector $\mathbf{Y}^{(2)}$.

Step 3. We now assign the trip-matrix to the network described by $\mathbf{Y}^{(2)}$ according to the route choice model and get new traffic flows $\mathbf{X}^{(3)}$. We then return to step 2 and iterate the steps 2 and 3 until the changes are no longer big enough to justify continuation of the process. Barbier and Haubrich state that the process converges very rapidly (after two or three iterations).

It is obvious that the order of succession of rerouting the x_{ij}'s in step 2 is important. It is also clear that an investment has a greater chance of being made for a large traffic flow than for a small one. And after an investment has been made we get the bundling effect that small flows will use the link which has been invested in, while the investment would not be made for, and the link would not be used by, the small flow alone. For this reason Barbier proposes to handle the x_{ij}'s in order of diminishing magnitude.

We will illustrate the operation of the method of Barbier and some of its weak points by a small example. Let us take the network of Figure (4.4.3).

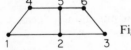

Figure 4.4.3   Network for the example of the method of Barbier

Suppose road 123 is a new road, which may be constructed (12, 21, 23 and $23 \in L_I$) and the other roads are old ones in which no investment need be made. Let us suppose a trip-matrix consisting of one relation: x^{13}. In the first step of the process the relation x^{13} will use the route 123 (minimum users' costs). In the second step we try to reroute $x_{12}^{(1)}$ and $x_{23}^{(1)}$. Suppose the investment costs for 12 and 23 do not outweigh the profits in users' costs made by using the new road. The result is that $x_{12}^{(1)}$ uses the route 1452 in the second step and $x_{23}^{(1)}$ uses 2563. At the same time we eject links 12 and 23 ($y_{12}^{(2)} = 0$ and $y_{23}^{(2)} = 0$). In the third step we assign x^{13} to the new network. The relation 13 will now use the route 14563. We see that the loop 525 introduced in the second step vanishes. For the flows $x_{14}^{(3)}$, $x_{45}^{(3)}$, $x_{56}^{(3)}$ and $x_{13}^{(3)}$ no better routes exist and the process is terminated.

One of the weak points of the Barbier's method can be illustrated by means of this example too. It is possible, of course, that the use of route 12 is better than the use of 1452 and the use of 23 is better than that of 2563 but that, on the contrary, the use of 14563 (without the loop 525) is better than the use of the new road 123. In this and similar cases Barbier's method does not find the correct optimal solution.

Barbier states that, in the cases where the optimal solution is not found, investment costs come out too high and so the users' costs are too 'low', a situation generally in favour of the users. Although the tendency does exist to minimize the users' costs 'more' than the investments this is not always the case. See, for instance, the network of Figure 4.4.4 in which we

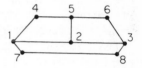

Figure 4.4.4   Network for the second example of the method of Barbier

assume 123 and 1783 to be new roads and 14563 and 52 old ones. In the first step route 123 is used, in the second step route 1452563, say, and in the third and last step route 14563, while 1783 might be the optimal route. With two, more or less parallel, new routes we can arrive at the situation where none of the new routes are used, whereas the use of one new route might be the optimal solution.

4.4.3.2   Haubrich's Extensions of the Method of Barbier

Haubrich (1972) has made some extensions to the original method of Barbier in his work on the railway networks for the Netherlands. The first extension, suggested already by Barbier himself, tries to decrease the effect mentioned at the end of the preceding section, namely that the whole can be better, while the parts are not. Therefore in the example of Figure 4.4.3 he does not try to reroute in the second step the flows $x_{12}^{(1)}$ and $x_{23}^{(1)}$ separately, but he tries to reroute $x_{123}^{(1)}$ all at once so avoiding the harmful effect of the loop 525. He accomplishes that by inspecting $x_{ij}^{(1)}$ and $x_{jk}^{(1)}$ and if $x_{ij}^{(1)} = x_{jk}^{(1)}$ he tries to reroute $x_{ij}^{(1)}$ and not $x_{ij}^{(1)}$ and $x_{jk}^{(1)}$ separately. Of course $x_{ij}^{(1)}$ and $x_{jm}^{(1)}$ must be inspected then for $l \neq i$ and $m \neq k$ and special precautions must be taken that these flows do not disturb the process.

The second extension is in the route choice criterion used in the second step. Barbier himself counts the investment costs if the link is not used by another flow and does not count any investment costs if the link is already used by another flow. Haubrich counts the investment costs *in toto* if the link is not used and counts them proportionally with the flows when the link is already used. So instead of relationship (4.4.15) we get:

find a path i to j such that:

$$\sum_{\substack{kl \in p \\ p \text{ forms a path from } i \text{ to } j}} \left\{ i_{kl} \frac{x_{ij}^{(1)}}{x_{ij}^{(1)} + x_{kl}^{(2)}} + t_{kl} x_{ij}^{(1)} \right\} \qquad (4.4.18)$$

is minimized.

The effect of this extension compared to the original method of Barbier is twofold. In the first place the system is less dependent on the order of succession of handling the $x_{ij}^{(1)}$'s. We may call this an advantage. In the second place it has become more difficult to reroute a flow because a part of the investment costs of other roads must always be paid. Thus the system will tend to retain direct connections and will benefit less from the effect of bundling flows. We might call this a disadvantage. It is difficult to say whether this extension is an improvement or not; that will depend, among other things, on the network, the trip-matrix and the goal of the study.

The third extension is based on the thought that in the rerouting process of the second step it is known that several flows will use a certain new link, but that this knowledge has been neglected in the first instance. So the flow to be rerouted has to 'pay' all the investment costs, although we know that the costs will be shared by other flows. To improve this situation Haubrich proposes to take into account, at the time of rerouting process, other possible users, known from the preceding step. So the criterion in relationship (4.4.15) becones now:

find a path from i to j such that:

$$\sum_{\substack{kl \in p \\ p \text{ forms a path from } i \text{ to } j}} \left\{ i_{kl} \frac{x_{ij}^{(1)}}{x_{ij}^{(1)} + x_{kl}^{(2)} + \alpha x_{kl}^{(1)}} + t_{kl} x_{ij}^{(1)} \right\} \quad \text{for } kl \neq ij \qquad (4.4.19)$$

or:

$$\left\{ i_{ij} \frac{x_{ij}^{(1)}}{x_{ij}^{(1)} + x_{ij}^{(2)}} + t_{ij} x_{ij}^{(1)} \right\}$$

is minimized.

In this method α has to be defined in advance, with $0 \leqslant \alpha \leqslant 1$ in general. Again it is difficult to say what improvement this extension brings about. It is difficult too to propose *a priori* a certain value of α. It too all depends on the network and the trip-matrix. Haubrich (1972) states however that he gets very good results with $\alpha = 0.25$, and, in fact, that he always finds the optimum for a number of test networks whereas the optimum has not been found for the same networks with other values of α (including $\alpha = 0$, the 'original' method of Barbier). Haubrich proposes some further extensions, which we will not deal with here.

4.4.4 Interactive Programming

Up to now we have supposed that the heuristic process is executed totally by the computer. Some features of human problem solving methods are incorporated into the computer program in advance and then the computer executes the process. Of these features of human problem solving, as mentioned in Section 4.4.1, some are very suitable for utilization by the computer but others are much better carried out by the human brains themselves. Of the latter we mention:

splitting the complex main problem into easier subproblems;
deciding to stop or to continue the searching process;
rejecting immediately solutions or sets of solutions that are not-promising.

On the other hand the computer is much to be preferred for the computation of the objective function or an upper or lower bound for it for a particular solution or set of solutions.

In view of this it would appear to be very profitable to combine the human brain and the computer for solving the problem in an interactive process. This has been proposed for instance by Stairs (1967) and Loubal (1967). Koike (1970) developed an interactive system in a theoretical paper. In this system, links can be added to and deleted from the network, both by man and by computer, as the computer evaluates the new networks and proposes further improvements according to certain search rules. Also at the Delft University of Technology work is being done along these lines (Jansen and Bovy, 1971), where a visual display and a light pen are used.

To illustrate the advantages of such an interactive system we cite Stairs 1967, page 226):

'The user (of an interactive system) is not just a cheap, non-linear, mixed-integer, optimizing element. He knows the weaknesses in the model and he may wish to change the model as the calculation proceeds, in order that new solutions may be considered or old ones excluded. When using the system to select a network he may wish to discard projects as soon as their worthlessness becomes apparent and he may be stimulated to invent projects not initially envisaged. The heuristic search rule and the interactive designer are not competing methods. Interactive trials may be the easiest way of developing effective search rules; on the other hand, heuristic rules may govern all the low-level searching while the designer grapples with the high-level strategy'.

To get a well-working interactive system it is necessary to use the computer in a real conversational manner. That means that it must be very easy to read input data and also to change the program (if possible) and that the results of the computation must be available in a very short time. All third generation computer systems offer these possibilities in principle.

4.5 HIERARCHICAL STRUCTURE, AGGREGATION AND DECOMPOSITION

We have seen that the size of the problem is the crucial difficulty in the optimization of realistic transport networks. Making use of the hierarchical structure of the transport network optimization problem it is possible to reduce the size of the problem by aggregation and/or decomposition. Aggregation and decomposition or partitioning are two general techniques for tackling large problems.

4.5.1 Hierarchical Structure

As we have seen, everything depends on everything in a transport network. So everything must be considered simultaneously. However, closer analysis shows that some variables or sets of variables are more strongly interrelated than others. Moreover we see that there are variables of which the values have impacts on many other variables, while other variables only have impacts on smaller subsets of variables, while others again have almost no impact at all and are only influenced themselves by the values of other variables.

From this we may conclude that there exist some hierarchies in a transport network. Some links and nodes serve as 'first-level' links and nodes: they have impacts on many other links and nodes; the next group 'second-level' links and nodes depends on the first group and influences, for its own part, other links and nodes, but not to the same extent as the first group. Links and nodes of a third, fourth and so on level may exist, with still less impact on other links and nodes.

In a road network we may consider the most important highways forming the first-level network, the remaining highways forming the second-level network, the secondary roads forming the third level and so on. In a public transport network we can make the same distinction. Moreover we may consider the location of the stops and the operating-schemes to be on a lower level than the structure and dimensions of the public transport network itself.

In addition to the different levels to be distinguished in the transport network the optimization problem can also be seen to be hierarchically structured (see for instance Manheim, 1966). At the highest level only the main lines are concerned and the question is the defining of fields of interest. At this first level many things are uncertain. The second level goes a little deeper: more details are considered, though not very accurately, and there is some more certainty. The next level is again more detailed and certain and so on. Working in this way, first very rough research is done, just to know where and what to investigate, while next rough estimates of possible net-

works are made and so on until, having worked through a whole series of
levels, at last some very detailed networks are proposed and evaluated.

4.5.2 Aggregation

Aggregation is an application of the hierarchical structure of a transport
network. In aggregation some variables are put together into one new variable to reduce the size of the problem; the mere omission of less relevant
variables can also be called aggregation. This aggregation can be done—
and always is done—manually by the problem-solver at the stage of the
problem formulation, but it can also be done by the computer with the use
of an aggregation-algorithm. There are two important types of aggregation
in transport networks:

(a)   zone aggregation;
(b)   link aggregation.

Zone aggregation means that we do not assume that the traveller can
travel from every possible origin to every possible destination but only from
and to the centroids of transport zones. This is always done in a transport
study (see for instance Overgaard, 1966, Wohl and Martin, 1969 or Hamerslag, 1972). In a narrower sense, zone aggregation means the aggregation
of some transport zones into one super zone with one new centroid.

Link aggregation means that we do not include all existing or possible
roads as links in the network but only a smaller number. There are two main
principles (Chan et al., 1968):

(a)   link extraction;
(b)   link abstraction.

Link extraction means that we just omit the unimportant links. This is the
most popular form of link aggregation. If we do this we must also omit the
trips which we assume would have used the omitted links. This can be done,
for instance, by omitting the short-distance trips or the diagonal of the trip-
matrix, assuming that these trips would have made use of the omitted
second-level links. Link abstraction also means that we use a small number
of links (smaller than the real number) but that we do not just omit some
links. Instead we incorporate some properties of a set of links into one
aggregation link. For instance, we replace three two-lane roads by one link
with the capacity of three two-lane roads and the speed-flow relationships
of a two-lane road.

Chan (1969) uses an aggregation algorithm in which zone aggregation
and link abstraction are accomplished according to certain criteria. He
then generally uses the aggregate network to compute lower and upper
bounds in his branch and bound algorithm (see Section 4.3.4). But sometimes
he returns to the original network to compute stronger bounds and then he
also obtains new values for the variables in the aggregate network.

4.5.3 Decomposition or Partitioning

Decomposition or partitioning is used to split one very large problem into several smaller ones of manageable size. It is possible to split the problem into several equivalent problems or into a master problem with a number of subproblems. This last type of decomposition is an especially typical application of the hierarchical structure of the problem. The splitting of a large problem into smaller ones and focusing on these is, of course, always done implicitly in research (the whole universe is never studied in one project), but we will treat this technique explicitly here.

The decomposition can be best carried out when the objective function is separable and when it is possible to split the constraints into sets containing only variables contained in the parts of the separated objective function and a set of common constraints:

$$\min_{\mathbf{X}_1, \mathbf{X}_2, \dots \mathbf{X}_n} F(\mathbf{X}_1, \mathbf{X}_2, \dots \mathbf{X}_n) = \sum_{i=1}^{n} F_i(\mathbf{X}_i)$$

subject to $\mathbf{G}_i(\mathbf{X}_i) >, =, <0$ with $i = 1, \dots, n$

$$\mathbf{H}(\mathbf{X}_1, \mathbf{X}_2, \dots \mathbf{X}_n) >, =, <0$$

(4.5.1)

In the decomposition we try to solve independently n problems, closely related to the optimization of the elements of the separated objective function and corresponding sets of constraints, while an overall policy takes care of the common constraint:

$$\min_{\mathbf{X}_i} F'(\mathbf{X}_i) \quad \text{subject to } \mathbf{G}_i(\mathbf{X}_i) >, =, <0 \left. \begin{array}{c} \\ \\ \end{array} \right\} \quad i = 1, \dots, n$$

$$\mathbf{H}_i(\mathbf{X}_i) >, =, <0$$

+ an overall policy

The decomposition has to be made manually by the problem-solver, but it must be possible too to construct algorithms to do the job according to certain criteria.

For the network optimization there are two important ways of partitioning or decomposition:
(a) geographic partitioning;
(b) pure hierarchic partitioning.

In geographic partitioning we divide the study area into geographic units and optimize the networks for them, while the master plan takes care of the overall optimal network. In pure hierarchic partitioning we divide the network into primary roads, secondary roads and so on and optimize those network systems separately, while the master plan takes care of the whole. Of course a combination of geographic and pure hierarchic partitioning can also be profitable.

Linear programs are always in the form of relation (4.5.1), so decomposition is always applicable, especially when the sets of constraints G_i are much larger than the set H. So it is obvious that there exist standard principles for the decomposition of linear programs (Dantzig and Wolfe, 1960). In the Dantzig–Wolfe decomposition principle, optimal solutions are sought for the subproblems i subject to the constraints G_i as the objective function F_i is changed into F_i', such that the common constraint H is taken into account. The master program combines the feasible partial solutions into an overall solution such that the overall objective function is minimized and the constraint H is also met.

Some methods, which will be discussed in the following chapters, such as the method of the stepwise assignment according to the least marginal objective function and dynamic programming are also examples of decompositions. Of course combinations of several decompositions (say dynamic programming combined with geographic and pure hierarchic partitioning) are also possible and are probably even very profitable.

REFERENCES

Barbier, M. (1966). Le future réseau de transports en Région de Paris. *Cahiers de l'Institut d'Aménagement et d'Urbanisme de la Région Parisienne*, **4–5**, No. 4.

Bergendahl, G. (1969). *Models for Investments in a Road Network*. Bonniers, Stockholm, 1969.

Bergendahl, G. (1971). *Principles of Heuristic Programming*. Research Report 62, Department of Business Administration, Stockholm University, Stockholm.

Bhatt, K. U. (1968). *Search in Transportation Planning: A Critical Bibliography*, Research Report R68-46, M.I.T. Department of Civil Engineering, Cambridge, Mass.

Boyce, D. E., and Farhi, A. (1972). *Réseaux de Transport Optimaux*. Paper presented at the First Congress of the Regional Sciences Association of North-Western Europe, Rotterdam.

Bruynooghe, M. (1972). An optimal method of choice of investments in a transport network. Paper presented at the Planning & Transport Research & Computation Seminars on *Urban Traffic Model Research*, London (8–12 May).

Bureau Central d'Etudes pour les Equipements d'Outre-Mer (1967). *Etude Méthodologique pour la Recherche d'une Sequence Optimale d'Investissements*, Paris.

Carter, E. C., and Stowers, J. R. (1963). *Model for Funds Allocation for Urban Highway Systems Capacity Improvements*. Highway Research Board Record, No. 20.

Chan, Y-P., Follansbee, K. G., Manheim, M. L., and Mumford, J. (1968). *Aggregation in Transport Networks: An Application of Hierarchical Structure*. Research Report R68-47, M.I.T., Department of Civil Engineering, Cambridge, Mass.

Chan, Y-P. (1969). *Optimal Travel Time Reduction in a Transport Network: An Application of Network Aggregation and Branch and Bound Techniques*. Research Report R69-39, M.I.T., Department of Civil Engineering, Cambridge, Mass.

Dantzig, G. B., and Wolfe, Ph. (1960). Decomposition principle for linear programs. *Operations Research*, **8**, No. 1.

Gilmore, P. C. (1962). Optimal and suboptimal algorithms for the quadratic assignment problem. *Journal of the Society for Industrial and Applied Mathematics*, **10**, 305.

Goldman, A. J., and Nemhauser, G. L. (1967). A transport improvement problem transformable to a best-path problem. *Transportation Science*, 1, 295–307.

Golomb, S. W., and Baumert, L. D. (1965). Backtrack programming. *Journal of the Association of Computing Machinery*, **12**, 516.

Gomory, R. E., and Hu, T. C. (1964). Synthesis of a communication network. *Journal of the Society for Industrial and Applied Mathematics*, **12**, No. 2 (June).

Hadley, G. (1962). *Linear Programming*, Addison-Wesley, Reading, Mass.

Hadley, G. (1964). *Nonlinear and Dynamic Programming*, Addison-Wesley, Reading, Mass.

Hamerslag, R. (1972). *Prognosemodel voor het Personenvervoer in Nederland*. Koninklijke Nederlandse Toeristenbond A.N.W.B., 's-Gravenhage.

Haubrich, G.Th.M. (1972). De optimalisering van het spoorwegnet in Nederland ten behoeve van het personenvervoer, *Tijdschrift voor Vervoerswetenschap* (extra number).

Hershdorfer, A. M. (1965). *Optimal Utilisation and Synthesis of Road Networks*, M.I.T., Department of Civil Engineering, Cambridge, Mass.

Hu, T. C. (1969). *Integer Programming and Network Flows*, Addison-Wesley, Reading, Mass.

Jansen, G. R. M. (1971). *Transport Network Optimization: A Preliminary Bibliography*, 2nd edition. Working Paper No. OTN/3/71.4, Transportation Research Laboratory, Delft University of Technology, Delft.

Jansen, G. R. M., and Bovy, P. H. L. (1971). Verkeers- en vervoersresearch aan de TH Delft 1970, *Verkeerstechniek*, **22**, No. 3 (March).

Koike, H. (1970). *Planning Urban Transportation Systems: A Model for Generating Socially Desirable Transportation Network Configurations*. Urban Transportation Program Research Report 2, Department of Urban Planning and Civil Engineering, University of Washington, Seattle.

Land, A. H., and Doig, A. (1960). An automatic method for solving discrete programming problems. *Econometrica*, **28**, 497.

Lawler, E. L., and Wood, D. E. (1966). Brach-and-bound methods: a survey. *Operations Research*, **14**, 699.

Loubal, P. S. (1967). *A Network Evaluation Procedure*, Highway Research Record 205.

Manheim, M. L. (1966). *Hierarchical Structure: A Model of Design and Planning Processes*, M.I.T. Press, Cambridge, Mass.

Manheim, M. L., with Ruiter, E.R., and Bhatt, K.U. (1968). *Search and Choice in Transport Systems Planning: Summary Report*, Research Report R68-40; M.I.T., Department of Civil Engineering, Cambridge, Mass.

Mitten, L. G. (1970). Branch-and-bound Methods: General Formulation and Properties, *Operations Research*, **21**, No. 1 (January/February).

Nederlands Economisch Intituut (1972). *Integrale Verkeers- en Vervoerstudie (annex VI)*, Staatsuitgeverij, s'Gravenhage.

Ochoa-Rosso, F. (1968). *Applications of Discrete Optimization Techniques to Capital Investment and Network Synthesis Problems*. Research Report R68-42, M.I.T., Department of Civil Engineering, Cambridge, Mass.

Ochoa-Rosso, F., and Silva, A. (1968). *Optimum Project Addition in Urban Transportation Networks via Descriptive Traffic Assignment Models*, Research Report R 68-44, M.I.T., Department of Civil Engineering, Cambridge, Mass.

Overgaard, K. R. (1966). *Traffic Estimation in Urban Transportation Planning*, Acta Polytechnica Scandinavia.

Pearman, A. (1971). The choice of projects from among a set of proposed improvements to a network. Paper presented at the Planning & Transport Research & Computation Symposium on *Cost Models and Optimisation in Road Location, Design and Construction*, London (June).

Perret, F. (1967). *Strategies de Réalisation d'un Ensemble d'Operation*. Ministre de l'Equipement et du Logement Service des Affaires Economiques et Internationales, Paris.

Pothorst, R. (1971). *Een Wegnnetkonstruktie Probleem*. Mathematisch Centrum, Universiteit van Amsterdam, Amsterdam.

Ridley, T. M. (1965). *An Investment Policy to Reduce Travel Time in a Transportation Network*. Operation Research Centre Report ORC 65-34, University of California, Berkeley, Cal.

Ridley, T. M. (1968). An investment policy to reduce the travel time in a transportation network. *Transportation Research*, 2, No. 4 (December).

Roberts, P. O., and Funk, M. L. (1964). *Toward Optimum Methods of Link Addition in Transportation Networks*, M.I.T., Department of Civil Engineering, Cambridge, Mass.

Scott, A. J. (1967). A programming model of an integrated transportation network, *The Regional Science Association Papers*, 19.

Scott, A. J. (1969a). The optimal network problem: some computational procedures, *Transportation Research*, 3, No. 2 (July).

Scott, A. J. (1969b). Combinatorial programming and the planning of urban and regional systems. *Environment and Planning*, 1, 125.

Scott, A. J. (1971). *Combinatorial Programming, Spatial Analysis and Planning*, Methuen, London.

Stairs, S. (1968). Selecting an optimal traffic network. *Journal of Transport Economics and Policy*, 2, 218.

Steenbrink, P. A. (1970). *Optimalisering van de infrastruktuur: Notitie 1. Eerste resultaten van de Toespassing van de Gradiëntmethoden*, Internal note of the Nederlandse Spoorwegen, OP2/239/41(160), Utrecht.

Steenbrink, P. A. (1971). Optimalisering van de infrastruktuur, *Verkeerstechniek*, 22, No. 7 (July).

Tomlin, J. A. (1966). Minimum-cost multicommodity network flows, *Operations Research*, 14, No. 1 (January, February).

Ventker, R. (1970). *Die Ökonomischen Grundlagen der Verkehrsnetzplanung*, Verkehrswissenschaftliche Studien 11. Aus dem Institut für Verkehrswissenschaft der Universität Hamburg. Herausgegeben von H. Jürgensen und H. Diederich. Vandenhoeck und Ruprecht, Göttingen.

Villé, J. A. (1969). Model of investment optimization on an urban transportation network (O.P.T.R.A.). Proceedings of the Planning & Transport Research & Computation Symposium on *Cost Models and Optimisation in Road Location, Design and Construction*, London (June).

Wohl, M., and Martin, B. V. (1969). *Traffic Systems Analysis for Engineers and Planners*, McGraw-Hill, New York.

5

Stepwise Assignment According to the Least Marginal Objective Function: A New Method

5.1 INTRODUCTION

In the search for solution methods for the transport network optimization problem which can manage realistic networks, we have devised a new method: the stepwise assignment according to the least marginal objective function (Steenbrink, 1970, 1971). The most important feature of this method is the very short computation time needed. Only one complete assignment of the trip-matrix to the network needs to be made. So it is possible to handle very large networks in a reasonable computation time using this method. The method is, in the first place, suitable for costs minimization with a normative assignment. The possibilities for descriptive cases and surplus maximization are discussed in the Sections 5.5 and 5.6 respectively. In the stepwise assignment according to the least marginal objective function the original problem is decomposed into a master problem and a number of subproblems. The subproblems give for each link the optimal relationship between the dimension of the link and the traffic flow. In the master problem the traffic flows on the optimally-dimensioned network are chosen in such a way that the objective function is minimized (and the constraints are satisfied).

For convex functions it is possible to state the necessary and sufficient conditions for the optimal solution of the master problem. We try to reach this optimum by a stepwise assignment of the trip-matrix to the network according to the routes with the least value for the marginal objective function.

It is shown that both for convex and non-convex functions in this way reasonable solutions will be obtained. This method has been used for the optimization of the road networks in the Dutch Integral Transportation Study (see Part II of this book and Nederlands Economisch Instituut, 1972, annex V) for 351 origins and destinations and a network consisting of about 2,000 nodes and 6,000 links.

5.2 DECOMPOSITION OF THE PROBLEM

The problem of costs minimization is formulated as follows:

$$\min_{c_{ij}, x_{ij}} F = \sum_{ij \in L} F_{ij}(x_{ij}, c_{ij}) \qquad (5.2.1)$$

subject to:

the network constraints $\left.\vphantom{\begin{array}{c}a\\b\end{array}}\right\}$ $(A\mathbf{X} = \mathbf{0}, \mathbf{X} \geqslant \mathbf{0})$

a fixed trip-matrix

$$c_{ij}^{\min} \leqslant c_{ij} \leqslant c_{ij}^{\max} \quad \text{for all } ij \in L$$

or more precisely:

$$\min_{c_{ij}, x_{ij}^{ab}} F = \sum_{ij \in L} F_{ij}\left(\sum_{ab \in P} x_{ij}^{ab}, c_{ij} \right) \qquad (5.2.2)$$

subject to:

$$\sum_{\substack{i \\ (ij \in L)}} x_{ij}^{ab} - \sum_{\substack{k \\ (jk \in L)}} x_{jk}^{ab} = \begin{cases} = 0; & \text{for all } j \neq a \text{ or } b; j \in N \\ = -x^{ab} & \text{if } j = a \\ = x^{ab} & \text{if } j = b \end{cases} \quad \text{for all } ab \in P$$

$$x_{ij}^{ab} \geqslant 0 \qquad\qquad\qquad \text{for all } ij \in L; \text{ for all } ab \in P$$

$$c_{ij}^{\min} \leqslant c_{ij} \leqslant c_{ij}^{\max}; \qquad\qquad\qquad \text{for all } ij \in L$$

We decompose this problem into a master problem and n_L subproblems: (using the first formulation (5.2.1) of the problem).

The subproblems:

$$\left. \begin{array}{l} \min_{c_{ij}} F_{ij}(x_{ij}, c_{ij}) \\ \text{subject to} \quad c_{ij}^{\min} \leqslant c_{ij} \leqslant c_{ij}^{\max} \end{array} \right\} \quad \text{for all } ij \in L \qquad (5.2.3)$$

The solutions of this n_L problems give n_L relationships between the minimum value for the objective function per link and the traffic flow on that link (meeting the dimension constraint):

$$F_{ij}^{\min}(x_{ij}) = \min_{\substack{c_{ij} \\ (c_{ij}^{\min} \leqslant c_{ij} \leqslant c_{ij}^{\max})}} F_{ij}(x_{ij}, c_{ij}).$$

The master problem:

$$\min_{x_{ij}} \sum_{ij \in L} F_{ij}^{\min}(x_{ij}) \qquad (5.2.4)$$

subject to: $A\mathbf{X} = \mathbf{0}, \mathbf{X} \geqslant \mathbf{0}$

First we prove that the combination of the master problem and the sub-problems has the same solution as the original problem. Let us suppose F^* to be the optimal solution to the master problem (5.2.4). Then, of course, the following relationship holds:

$$F^* \leqslant \sum_{ij \in L} F_{ij}^{\min}(x_{ij})$$

for all $\mathbf{X}$ subject to $A\mathbf{X} = \mathbf{0}, \mathbf{X} \geqslant \mathbf{0}$.

From the statement of the subproblems (5.2.3) however it follows that for any particular solution $\mathbf{X}'$ holds:

$$\sum_{ij \in L} F_{ij}^{\min}(x'_{ij}) \leqslant \sum_{ij \in L} F_{ij}(x'_{ij}, c_{ij})$$

for all c_{ij} subject to $c_{ij}^{\min} \leqslant c_{ij} \leqslant c_{ij}^{\max}$.

Combining these two relationships it is seen that F^* is the minimum solution to the total problem and the decomposition is correct.

The subproblems (5.2.3) of defining the optimal relationship between the dimension and the traffic flow for each link are mathematically very easy to solve. The inspection of all possibilities, combined perhaps with a search method, or differentiating if we assume $F_{ij}(c_{ij})$ a differentiable function of the continuous variable c_{ij}, will yield the right solutions (see also Section 10.1). The master problem is mathematically much more complex. We will deal with it in the next two sections.

5.3 CONDITIONS FOR THE OPTIMAL SOLUTION OF THE MASTER PROBLEM

One of the chief difficulties in solving the master problem is the enormous number of variables and constraints for these variables. But fortunately we have met a similar problem before: the normative assignment problem of Section 2.2.2 has the same mathematical formulation. There, and also in Section 2.1.3, we showed, following Beckmann *et al* (1956) and others, that for convex functions we can derive conditions for the optimal solution that can be considered as conditions describing a state of equilibrium in trans-portation. Following the same approach as in those sections we will first derive the necessary and sufficient conditions for the optimal solution for the costs minimization problem.

The problem is stated as follows, using the notion of paths:

$$\min_{x_{ij}^{pab}} F(\mathbf{X}) = \sum_{ij \in L} F_{ij}^{\min}\left(\sum_{\substack{p \in Pa^{ab} \\ ab \in P}} x_{ij}^{pab} \right) \qquad (5.3.1)$$

subject to:

$$x_{ij}^{pab} = x_{kl}^{pab}; \qquad \text{for all } ij \in p \text{ and } kl \in p$$
$$\text{for all } p \in Pa^{ab} \qquad (5.3.2)$$
$$\text{for all } ab \in P$$

$$\sum_{p \in Pa^{ab}} x_{aj}^{pab} = x^{ab}; \qquad \text{for all } ab \in P \qquad (5.3.3)$$

$$x_{ij}^{pab} \geqslant 0; \qquad \text{for all } ij \in L$$
$$\text{for all } p \in Pa^{ab} \qquad (5.3.4)$$
$$\text{for all } ab \in P$$

Suppose $\mathbf{X}^*$ is the optimal solution, then the following relationship holds:

$$F(\mathbf{X}^*) \leqslant F(\mathbf{X}^* + \Delta\mathbf{X}) \qquad (5.3.5)$$

in which $\mathbf{X}^* + \Delta\mathbf{X}$ (with elements $(x_{ij}^{*\,pab} + \Delta x_{ij}^{pab})$) is a feasible solution in the 'neighbourhood' of $\mathbf{X}^*$. Because $\mathbf{X}^* + \Delta\mathbf{X}$ is feasible the following relationships hold for Δx_{ij}^{pab}:

from relationship (5.3.2):

$$\Delta x_{ij}^{pab} = \Delta x_{kl}^{pab}; \qquad \text{for all } ij \in p \text{ and } kl \in p$$
$$\text{for all } p \in Pa^{ab} \qquad (5.3.6)$$
$$\text{for all } ab \in P$$

from relationship (5.3.3):

$$\sum_{p \in Pa^{ab}} \Delta x_{aj}^{pab} = 0; \qquad \text{for all } ab \in P \qquad (5.3.7)$$

from relationship (5.3.4):

$$\Delta x_{ij}^{pab} \geqslant 0 \quad \text{if } x_{ij}^{*\,pab} = 0; \quad \text{for all } ij \in L$$
$$\text{for all } p \in Pa^{ab} \qquad (5.3.8)$$
$$\text{for all } ab \in P$$

Expanding $F(\mathbf{X}^* + \Delta\mathbf{X})$ into a Taylor series we get:

$$F(\mathbf{X}^* + \Delta\mathbf{X}) = F(\mathbf{X}^*) + \sum_{\substack{p \in Pa^{ab} \\ ab \in P \\ ij \in L}} \left(\frac{\partial F}{\partial x_{ij}^{pab}}\right)_{\mathbf{X} = \mathbf{X}^*} \Delta x_{ij}^{pab} + \text{higher order terms}$$
$$(5.3.9)$$

If F is a convex function of $\mathbf{X}$, then the sum of higher order terms is always positive. Let us suppose further that the values of the Δx_{ij}^{pab} are so small that the sum of the higher order terms is negligible (but still positive). That means that we can write the inequality (5.3.5) as:

$$\sum_{\substack{p \in Pa^{ab} \\ ab \in P \\ ij \in L}} \left(\frac{\partial F}{\partial x_{ij}^{pab}} \right)_{X = X^*} \Delta x_{ij}^{pab} \geqslant 0 \qquad (5.3.10)$$

As can be very easily seen we can write $(dF_{ij}^{\min}/dx_{ij})$ for $(\partial F/\partial x_{ij}^{pab})$. Let us further for brevity write F_{ij}' for the differential quotient in $X = X^*$. We then get:

$$\sum_{\substack{p \in Pa^{ab} \\ ab \in P \\ ij \in L}} F_{ij}' \, \Delta x_{ij}^{pab} \geqslant 0 \qquad (5.3.11)$$

We now substitute the relationships (5.3.6), (5.3.7) and (5.3.8) into (5.3.11). Substituting relationship (5.3.6) we get:

$$\sum_{\substack{p \in Pa^{ab} \\ ab \in P}} \Delta x_{ak}^{pab} \left(\sum_{ij \in p} F_{ij}' \right) \geqslant 0 \qquad (5.3.12)$$

We take now for all relations ab one 'free' variable Δx_{al}^{qab}, that means that $x_{al}^{*\,qab} > 0$, so that relationships (5.3.8) is not relevant and we derive from relationship (5.3.7):

$$\Delta x_{al}^{qab} = - \sum_{p \in Pa^{ab} - q} \Delta x_{ak}^{pab} \qquad (5.3.13)$$

Substituting relationship (5.3.13) into (5.3.12) we get:

$$\sum_{ab \in P} \left[\sum_{p \in Pa^{ab} - q} \left\{ \Delta x_{ak}^{pab} \left(\sum_{ij \in p} F_{ij}' \right) \right\} - \sum_{p \in Pa^{ab} - q} \Delta x_{ak}^{pab} \sum_{ij \in q} F_{ij}' \right] \geqslant 0$$

or:

$$\sum_{ab \in P} \sum_{p \in Pa^{ab} - q} \Delta x_{ak}^{pab} \left(\sum_{ij \in p} F_{ij}' - \sum_{ij \in q} F_{ij}' \right) \geqslant 0 \qquad (5.3.14)$$

Combining relationships (5.3.14) and (5.3.8) we get:

if $x_{ak}^{*\,pab} = 0$ then $\Delta x_{ak}^{pab} \geqslant 0$,

$$\left.\begin{array}{c}
\text{so it is a necessary condition that:} \\[4pt]
\displaystyle \sum_{ij \in p} F_{ij}' > \sum_{ij \in q} F_{ij}' \\[12pt]
\text{if } x_{ak}^{*\,pab} > 0 \quad \text{then } \Delta x_{ak}^{pab} \text{ is free,} \\[8pt]
\text{so it is a necessary condition that:} \\[4pt]
\displaystyle \sum_{ij \in p} F_{ij}' = \sum_{ij \in q} F_{ij}'
\end{array}\right\} \qquad (5.3.15)$$

From relationships (5.3.15) we observe first that q forms the 'shortest' (shortest in terms of the marginal objective function) route from a to b:

$$\sum_{ij \in p} F'_{ij} \geqslant \sum_{ij \in q} F'_{ij} = F'^{ab} \qquad (5.3.16)$$

With this knowledge we can write relationship (5.3.15) combined with the fact that $x_{ak}^{*\,qab} > 0$ and with relationship (5.3.2) as:

$$\left.\begin{array}{ll} \text{if } x_{ij}^{ab} > 0 \quad \text{then } F'^{ai} + F'_{ij} = F'^{aj} & \text{for all } ab \in P \\ \text{if } F'^{ai} + F'_{ij} > F'^{aj} \quad \text{then } x_{ij}^{ab} = 0 & \text{for all } ij \in L \end{array}\right\} \qquad (5.3.17)$$

From the convexity of $F(\mathbf{X})$ we know that the necessary conditions (5.3.17) are sufficient too and that the minimum is a global one. So we know that we can describe the solution for the costs minimization as a state in which for every relation only the routes with the least marginal costs are used. These conditions hold explicitly only for convex functions $F_{ij}^{\min}(x_{ij})$. For non-convex functions these conditions do not always describe a minimum solution.

5.4 SOLUTION METHODS FOR THE MASTER PROBLEM

5.4.1 Stepwise Assignment

Having the necessary and sufficient conditions for the optimal solution (for convex functions, anyway) the next question is what methods to use to achieve that optimal solution. Here too we make use of the resemblance with the assignment problem. For every relation the 'shortest' routes are used (in terms of $dF^{\min}/dx$) and furthermore this $dF^{\min}/dx$ is a non-negative increasing function of x as t is. So it can be easily understood that we can use here the same methods as used for the traffic assignment problem. In Section 2.1.4.2 we have already said that the algorithm does not seem to exist which would achieve an exact solution for the assignment problem within a reasonable computation time. We referred there to the work of Bruynooghe (1967), Gibert (1968), Bruynooghe, Gibert and Sakarovitch (1968), Murchland (1969) and Dafermos and Sparrow (1969).

We will describe here the method of stepwise assignment which is comparable with the capacity restraint technique well known in traffic engineering. In this stepwise assignment every relation of the trip-matrix is assigned part by part to the network. The parts of the relations of the trip-matrix are at each step assigned to the routes with the minimal value for the marginal objective function, computed with the flows assigned in the preceding steps. So the process is executed as follows. Assign in the first step a part α_1 of each relation of the trip-matrix to the network along the route with the least value for:

$$\sum_{ij \in \text{route}} \left(\frac{dF_{ij}^{\min}}{dx_{ij}} \right)_{x_{ij}=0} \tag{5.4.1}$$

After this assignment every link has a traffic flow $x_{ij}^{\alpha_1}$ on it. In the second step a part α_2 of every relation of the trip-matrix is assigned along the route with minimum value for:

$$\sum_{ij \in \text{route}} \left(\frac{dF_{ij}^{\min}}{dx_{ij}} \right)_{x_{ij}=x_{ij}^{\alpha_1}} \tag{5.4.2}$$

This process is continued till the whole trip-matrix has been assigned to the network:

$$\sum_n \alpha_n = 1 \tag{5.4.3}$$

5.4.2   Correctness of the Solutions Obtained

It is obvious that the results of this stepwise assignment always satisfy the constraints: since in every step the partial flows assigned are conservative, non-negative and additive to each other, the same is true of the final flows, and because of equation (5.4.3) the trip-matrix constraint has been satisfied at the end of the process. The conditions for the optimum have still to be fulfilled.

For a trip-matrix consisting of one relation (ab) and a network in which the different paths used do not contain the same links it can be proved that an infinite number of infinitely small steps gives the correct solution. It must be proved that:

$$\text{if } x_{ij} > 0 \quad \text{then } F'^{ai} + F'_{ij} = F'^{aj} \tag{5.4.4}$$

$$\text{and if } F'^{ai} + F'_{ij} > F'^{aj} \quad \text{then } x_{ij} = 0 \tag{5.4.5}$$

(working with one relation we can write x_{ij} for x_{ij}^{ab}).

First we will prove relationship (5.4.4). For $x_{ij} > 0$ there are two possibilities for the partial flow $dx_{ij}^{\text{last step}}$ assigned in the last step:

(a) $dx_{ij}^{\text{last step}} > 0$
(b) $dx_{ij}^{\text{last step}} = 0$

In case (a) the link ij must be contained in the route with minimal value for the marginal objective function computed at the point $\mathbf{X} - d\mathbf{X}$ and also at the final assignment $\mathbf{X}$ if $d\mathbf{X}$ approaches zero and F' is a continuous function of $\mathbf{X}$.

In case (b) we can assume that ij is contained in path p that has been used at the point $\theta\mathbf{X}$, with $0 \leqslant \theta < 1$ and not afterwards. Moreover we can assume that, at the last step, path r is used. Then the following relationship holds:

using F'^p for $\sum\limits_{ij \in p} F'_{ij}$

$$(F'^p)_{\theta \mathbf{X}} \leqslant (F'^r)_{\theta \mathbf{X}} \leqslant (F'^r)_{\mathbf{X}-d\mathbf{X}} \leqslant (F'^p)_{\mathbf{X}-d\mathbf{X}} \qquad (5.4.6)$$

due to the assignment rule and the convexity of the objective function. But because path p has not been used after the point $\theta \mathbf{X}$ (5.4.7) also holds:

$$(F'^p)_{\mathbf{X}-d\mathbf{X}} = (F'^p)_{\theta \mathbf{X}+d\mathbf{X}} \qquad (5.4.7)$$

Combination relationships (5.4.6) and (5.4.7) gives

$$(F'^p)_{\theta \mathbf{X}} \leqslant (F'^r)_{\mathbf{X}-d\mathbf{X}} \leqslant (F'^p)_{\theta \mathbf{X}+d\mathbf{X}} \qquad (5.4.8)$$

Letting $d\mathbf{X}$ approach zero and assuming F to be a continuous function of $\mathbf{X}$ we see that:

$$(F'^r)_{\mathbf{X}} = (F'^p)_{\theta \mathbf{X}} = (F'^p)_{\mathbf{X}} \qquad (5.4.9)$$

So all paths used are shortest paths and relationship (5.4.4) has been proved.

The proof of relationship (5.4.5) is very simple. If link ij is not contained in the shortest path at the final state, it will certainly not be so contained before, because of the convexity of the objective function and the fact that the traffic flows on the different paths can only increase during the assignment process. Because link ij is never contained in the shortest path, it will never get any traffic flow, thus relationship (5.4.5) has been proved.

So we see that, in the case of one relation, a network in which the different paths used have no links in common and a convex objective function the stepwise assignment according to the least marginal objective function with an infinite number of infinitely small steps yields the correct optimal solution.

In cases with more relations or even with one relation when the different paths used have links in common, an infinite number of infinitely small steps will not always give the correct solution. Relationship (5.4.7) does not hold any longer. It is possible that a link shared by more paths is originally contained in a certain shortest path but is not at the final state because of traffic flows assigned to other paths but the same link. See Figure 5.4.1. Suppose there are two relations 12 and 34. Suppose further that the marginal objective function is rapidly increasing on link 56 and is high but slowly

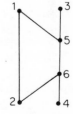

Figure 5.4.1   Network in which an infinite number of steps still does not give the correct solution

increasing on link 12. The optimal solution might be that relation 12 uses link 12 and relation 34 route 3564 while the stepwise assignment according to the least marginal objective function might result in a use of 12 and 1562 for relation 12.

So we see that relationship (5.4.4) is not always true. Relationship (5.4.5) is always true, as can easily be seen. It is rather difficult to say what the ultimate effect of these things is, but it is probably not too serious.

Still it is obvious that an infinite number of infinitely small steps will give better results than a smaller number, because in that case the optimum routes 'change more slowly' so that better routes can be chosen. For the parts assigned in the nth step the following conditions hold:

$$\left.\begin{array}{l} \text{if } \mathrm{d}x_{ij}^{\alpha_n ab} > 0 \quad \text{then } (F'^{ai})_{\mathbf{X}=\mathbf{X}^{\alpha_n-1}} + (F'_{ij})_{\mathbf{X}=\mathbf{X}^{\alpha_n-1}} = (F'^{aj})_{\mathbf{X}=\mathbf{X}^{\alpha_n-1}} \\ \text{if } (F'^{ai})_{\mathbf{X}=\mathbf{X}^{\alpha_n-1}} + (F'_{ij})_{\mathbf{X}=\mathbf{X}^{\alpha_n-1}} > (F'^{aj})_{\mathbf{X}=\mathbf{X}^{\alpha_n-1}} \quad \text{then } \mathrm{d}x_{ij}^{\alpha_n ab} = 0 \end{array}\right\} \quad (5.4.10)$$

Of course it would be better if the differential quotients were taken in $\mathbf{X}^{\alpha_n}$ instead of $\mathbf{X}^{\alpha_n-1}$. We have said already that it is therefore preferable to use many small steps so that $\mathbf{X}^{\alpha_n-1}$ approaches $\mathbf{X}^{\alpha_n}$.

We now look at the case of non-convex functions $F_{ij}^{\min}(x_{ij})$. In this case we cannot assume that the sum of the higher order terms of the Taylor expansion (5.3.9) of the objective function is always positive. This means that we cannot derive the same conditions for the optimal solution, which means again that the assignment according to the least marginal objective function does not guarantee a good solution.

It does not, however, mean that the method will always give a wrong solution in the case where not all functions are convex. To demonstrate this we will discuss a case in which the method also gives a correct solution for non convex functions. Consider two possible paths: path 1 and path 2. If the marginal objective function on one path is always smaller than on the other one, only that path will be used. This is also the correct solution, as will be proved below. Suppose:

$$\sum_{ij \,\in\, \text{path}\,1} \left(\frac{\mathrm{d}F_{ij}^{\min}}{\mathrm{d}x_{ij}}\right)_{x_{ij}=x_{ij}^{\alpha}} \leqslant \sum_{kl \,\in\, \text{path}\,2} \left(\frac{\mathrm{d}F_{kl}^{\min}}{\mathrm{d}x_{kl}}\right)_{x_{kl}=x_{kl}^{\beta}} \quad (5.4.11)$$

$$\text{for } x_{ij}^{\alpha} \geqslant x_{kl}^{\beta}$$

In this case only path 1 will be used. To prove that this is the correct optimal solution we have to prove that:

$$\sum_{ij \,\in\, \text{path}\,1} F_{ij}(x_{ij}^{\alpha}) \leqslant \sum_{ij \,\in\, \text{path}\,1} F_{ij}(x_{ij}^{\beta}) + \sum_{kl \,\in\, \text{path}\,2} F_{kl}(x_{kl}^{\gamma})$$

$$\text{with } x_{ij}^{\beta} + x_{kl}^{\gamma} = x_{ij}^{\alpha} \tag{5.4.12}$$

We can write relationship (5.4.12) as:

$$\sum_{ij \in \text{path 1}} \int_0^{x^\alpha} \frac{dF_{ij}}{dx_{ij}} dx_{ij} \leqslant \sum_{ij \in \text{path 1}} \int_0^{x^\beta} \frac{dF_{ij}}{dx_{ij}} dx_{ij} + \sum_{kl \in \text{path 2}} \int_0^{x^\gamma} \frac{dF_{kl}}{dx_{kl}} dx_{kl}$$

or

$$\sum_{ij \in \text{path 1}} \int_0^{x^\beta} \frac{dF_{ij}}{dx_{ij}} dx_{ij} + \sum_{ij \in \text{path 1}} \int_{x^\beta}^{x^\alpha} \frac{dF_{ij}}{dx_{ij}} dx_{ij}$$

$$\leqslant \sum_{ij \in \text{path 1}} \int_0^{x^\beta} \frac{dF_{ij}}{dx_{ij}} dx_{ij} + \sum_{kl \in \text{path 2}} \int_0^{x^\alpha - x^\beta} \frac{dF_{kl}}{dx_{kl}} dx_{kl}$$

(5.4.13)

But relationship (5.4.13) is, in fact, true because

$$\frac{dF_{ij}}{dx_{ij}} \leqslant \frac{dF_{kl}}{dx_{kl}} \quad \text{for } x_{ij} = x_{kl} + x^\beta$$

so the correct optimal solution is obtained.

In Figure 5.4.2 we give a few illustrations of functions $F_{ij}(x_{ij})$ on a network of two links for which the stepwise assignment according to the least marginal objective function does or does not give the correct optimal solution.

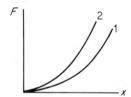

(a) Correct solution obtained (convex)

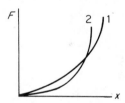

(b) Correct solution obtained (convex)

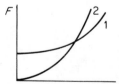

(c) Correct solution not always obtained
(F_1 is not convex in $x_1 = 0$)

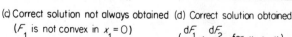

(d) Correct solution obtained
$\left(\dfrac{dF_1}{dx_1} \leq \dfrac{dF_2}{dx_2} \text{ for } x_1 \geq x_2 \right)$

Figure 5.4.2   Functions $F_{ij}(x_{ij})$ for a network of two links, for which the stepwise assignment according to the least marginal objective function does or does not give the correct optimal solution.

In Chapter 9 we will give a treatment of a 'real' relationship between $F_{ij}^{\min}$ and x_{ij} and there we show that for continuous c_{ij} the function has a form as in Figure 5.4.3 (see also the figures in Section 10.4). The switching

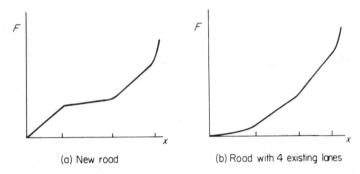

(a) New road   (b) Road with 4 existing lanes

Figure 5.4.3   Relationship between objective function and traffic flow

points correspond with 4, 8 and 12 lanes for a road. Here we see that the functions are convex except for new roads in the area up to 4 lanes. We have here assumed a continuous number of lanes. Working with an integral number of lanes we should get a continuous but less convex relationship between objective function and traffic flow, such as is shown in Figure 5.4.4 (see also Section 10.5).

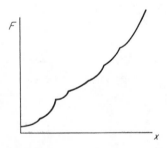

Figure 5.4.4   Relationship between objective function
and traffic flow for an integral number of lanes

A popular example of non-convex functions is the case in which the construction of one large connection is better than the construction of two small ones. In Figure 5.4.5 we show such a situation. We assume two relations 12 and 34 and for the relationship between objective function and traffic flow

Figure 5.4.5   Example of the case that one large road is better than
two small ones

the relationship of Figure 5.4.3(a). The 'trip-matrix' is such that in the optimal
solution the link 56 and the links 15, 35, 26 and 46 are constructed. Using
the stepwise assignment according to the least marginal objective function
in all cases the links 12 and 34 will be constructed. Only if the traffic flows
are very heavy the network with link 56 is constructed in addition. So the
stepwise assignment according to the least marginal objective function gives
a wrong solution in this case. Fortunately a better solution would be obtained
if the relations 14 and 32 existed, too, and certainly if there were a heavy
transport relation 56.

5.4.3   Parameters of the Method

We have seen that a stepwise assignment according to the least marginal
objective function will generally give reasonable solutions both for convex
and non-convex objective functions. For convex objective functions especially a large number of steps is desired. For the assignment in the nth step
the conditions of relationship (5.4.10) hold, which say that the next partial
flows are assigned to the routes with the least values for the marginal objective function computed from the flows resulting from the $(n - 1)$th step.
Of course it would be better to use the values of the marginal objective function of the same step. Therefore the use of many small steps is desired. On
the other hand the necessary computing time imposes strong upper bounds
on the number of steps. For every step a whole shortest path computation
and assignment procedure is needed, which takes a lot of computing time
(see Chapter 7). So we will try to use as few steps as possible.

It is advantageous, then, to use difference quotients instead of differential
quotients (this has been done in the Dutch Integral Transportation Study,
see Chapter 11). The criterion then becomes (ignoring division by the
constant Δx):

$$\{F_{ij}^{\min}(x_{ij}^{\alpha_{n-1}} + \Delta x) - F_{ij}^{\min}(x_{ij}^{\alpha_{n-1}})\} \tag{5.4.14}$$

For the first step this reduces to:

$$F_{ij}^{\min}(\Delta x)$$

(5.4.15)

The advantage of this is that we 'look a little ahead', dependent on the magnitude of Δx.

Another way of reducing the errors made when only a few steps are used is not to take:

$$(F')_{\mathbf{X} = \mathbf{X}^{\alpha_{n-1}}}$$

as criterion, but:

$$(F')_{\mathbf{X} = \mathbf{X}^{\beta_{n-1}}}$$

(5.4.16)

with $\mathbf{X}^{\beta_{n-1}} \geqslant \mathbf{X}^{\alpha_{n-1}}$.

This is accomplished by, in the first step, assigning a part α_1 and a part β_1 of the trip-matrix (with $\beta_1 \geqslant \alpha_1$) to the network according to the route with least value for:

$$\sum_{ij\,\in\,\text{route}} \left(\frac{\mathrm{d}F_{ij}}{\mathrm{d}x_{ij}}\right)_{x_{ij} = x_{ij}^{\beta_0}}$$

in which it is not necessary that $x_{ij}^{\beta_0} = 0$.

In the second step a part α_2 and a part β_2 are assigned to the network according to the route with lowest value for:

$$\sum_{ij\,\in\,\text{route}} \left(\frac{\mathrm{d}F_{ij}}{\mathrm{d}x_{ij}}\right)_{x_{ij} = x_{ij}^{\beta_1}}$$

Every link now gets two 'traffic flows':

$$x_{ij}^{\alpha_2} = x_{ij}^{\alpha_1} + \alpha_2 \sum_{\substack{ab \text{ uses a path in} \\ \text{which } ij \text{ is contained}}} x^{ab}$$

$$x_{ij}^{\beta_2} = x_{ij}^{\alpha_1} + \beta_2 \sum_{\substack{ab \text{ uses a path in} \\ \text{which } ij \text{ is contained}}} x^{ab}$$

The process is continued till the whole trip-matrix is definitely assigned:

$$\sum_n \alpha_n = 1$$

Of course it is not necessary, and hardly even possible that:

$$\sum_n \beta_n = 1.$$

It is obvious that it is possible to use a combination of the technique described last and the use of difference quotients.

Having the method (using $\{F_{ij}^{\min}(x_{ij} + \Delta x) - F_{ij}^{\min}(x_{ij})\}$ as the criterion) we have to define three assignment parameters:

(a) the number of steps;
(b) the magnitude of the parts to be assigned to the network at the different steps (α_n can be different for every n);
(c) the magnitude of Δx (this can also be different for every step, so Δx^n).

For convex functions $F_{ij}^{\min}(x_{ij})$ we have already said that theoretically the use of an infinite number of steps and, in connection with this, infinitely small values for α_n and Δx^n are preferred. Mindful of computing time we try to make the number of steps as small as possible. Of course we define the α_n and the Δx^n dependent on the number of steps (and on each other). In the capacity-restraint technique it is usual to reduce the size of the α_n with increasing n (for instance: 0·45, 0·25, 0·15, 0·10, 0·05, see Steel, 1965).

Knowing that not all functions may be convex for all values of x is one more reason for not using an infinite number of steps. Furthermore we may try to choose the parameters n, α_n and Δx^n in such a way that the stepwise assignment according to the least marginal objective function yields a good solution for the type of functions used. In Section 5.7 we will show some sensitivity analysis on a small testing network, while we will give a further treatment of the choice of the assignment parameters for a real network in Chapter 11.

5.5 THE USE OF A NORMATIVE ASSIGNMENT *VERSUS* A DESCRIPTIVE ONE AND THE PRICE MECHANISM

Here we used an assignment that minimizes the total social costs, so it is a special kind of normative assignment. The descriptive assignment will give different flows and will also give a different (and higher) value for the objective function and another network. Because a descriptive assignment describes reality (or tries to do so), the use of a normative assignment is certainly an objection. One possible way of reducing this objection is to start by defining the optimal network with the normative assignment, assigning the trip-matrix to the network obtained with a descriptive method and adjusting the network to the traffic flows (see Section 11.2.1). If desirable this process can be iterated. We get the method of the continuous optimal adjustment, as described in Section 4.4.2.1, with a fairly good starting network.

Another observation we can make about the use of the normative assignment is that the normative assignment gives the lowest value for the objective function and we may try to influence the individual route choice in such a way that the results of the real route choice behaviour are as near as possible to the normative assignment (compare Section 2.2.2.1). We can try to do that with such measures as traffic management and road pricing. Using a charging system it is easy to see how high the charges must be. For the normative assignment the 'route choice' criterion is:

$$\frac{\mathrm{d}F^{\min}}{\mathrm{d}x} \quad \text{(omitting the } ij\text{'s)} \tag{5.5.1}$$

Let us suppose the real behavioural route choice is described by individual users' costs minimization. So the criterion there is t. We must charge the difference between these two criteria in order to obtain the coincidence of the normative and the descriptive assignments. So the charge must be:

$$\frac{\mathrm{d}F^{\min}}{\mathrm{d}x} - t \tag{5.5.2}$$

These charges serve only to influence the route choice of the traveller. Yet it is interesting to look at the total money received:

$$\left(\frac{\mathrm{d}F^{\min}}{\mathrm{d}x} - t\right)x \tag{5.5.3}$$

The investment costs are:

$$i = F^{\min} - tx \tag{5.5.4}$$

Subtraction of relationship (5.5.4) from (5.5.3) gives the difference between the money received and the investment costs:

$$\frac{\mathrm{d}F^{\min}}{\mathrm{d}x}x - F^{\min} = \frac{\mathrm{d}F^{\min}}{\mathrm{d}x}x - \int_0^x \frac{\mathrm{d}F^{\min}}{\mathrm{d}\xi}\,\mathrm{d}\xi \tag{5.5.5}$$

When $F^{\min}(x)$ is a convex function we have:

$$\frac{\mathrm{d}F^{\min}}{\mathrm{d}x} \geqslant \frac{\mathrm{d}F^{\min}}{\mathrm{d}\xi} \quad \text{for } x \geqslant \xi.$$

So the difference between money received and investment costs is never negative.

At least this holds for convex functions. For concave functions (as in the case of small roads and of initial investments) this difference is, in fact, negative. This is the so-called 'fundamental deficit of infrastructure' (Allais et al., 1965, Oort, 1966).

5.6 THE USE OF THE METHOD FOR SURPLUS MAXIMIZATION

After this side-track into the price-mechanism we return to the method and ask if it is useful for surplus maximization. For reasons of simplicity we assume that an (inverse) demand function exists for every relation:

$$t^{ab} = g^{ab}(x^{ab}) \quad \text{for all } ab \in P \tag{2.1.4}$$

We can define the consumers' benefits of travelling in relation ab as:

$$\int_0^{x^{ab}} g^{ab}(x)\,dx \tag{3.2.4}$$

We state the normative surplus maximization problem (4.1.2) as follows (minimizing the negative of the total surplus):

$$\min_{c_{ij}, x_{ij}^{pab}} \left[\sum_{ij \in L} \left\{ \sum_{\substack{p \in Pa^{ab} \\ ab \in P}} x_{ij}^{pab} t_{ij}\left(\sum_{\substack{p \in Pa^{ab} \\ ab \in P}} x_{ij}^{pab}, c_{ij} \right) + i_{ij}(c_{ij}) \right\} \right.$$

$$\left. - \sum_{ab \in P} \int_0^{\sum\limits_{p \in Pa^{ab}} x_{ak}^{pab}} g^{ab}(x)\,dx \right]$$

subject to: the network constraints

$$c_{ij}^{\min} \leqslant c_{ij} \leqslant c_{ij}^{\max} \tag{5.6.1}$$

This problem can also be decomposed into a master problem and n_L subproblems. The subproblems are similar to the subproblems (5.2.3) for the costs minimization and they have the same solutions because the only difference between the objective function of problem (5.6.1) and the objective function for the costs minimization is the last term, which is constant with respect to the dimensions c_{ij}.

The master problem has the form:

$$\min_{x_{ij}^{pab}} \left\{ \sum_{ij \in L} F_{ij}^{\min}\left(\sum_{\substack{p \in Pa^{ab} \\ ab \in P}} x_{ij}^{pab} \right) - \sum_{ab \in P} \int_0^{\sum\limits_{p \in Pa^{ab}} x_{ak}^{pab}} g^{ab}(x)\,dx \right\}$$

subject to the network constraints   (5.6.2)

The second term is concave, so the negative is convex. Thus, when the first term is convex, the total objective function is convex and we can derive the necessary and sufficient conditions for the optimal solution:

$$\left. \begin{aligned} \text{if } x_{ij}^{pab} > 0 \quad \text{then} \quad -g^{ab} + \sum_{ij \in p} \frac{dF_{ij}^{\min}}{dx_{ij}} &= 0 \quad \text{for all } p \in Pa^{ab} \\[2mm] \text{if } -g^{ab} + \sum_{ij \in p} \frac{dF_{ij}^{\min}}{dx_{ij}} &> 0 \quad \text{then } x_{ij}^{pab} = 0 \quad \text{for all } ab \in P \end{aligned} \right\} \tag{5.6.3}$$

So the routes with least marginal objective function must be 'used' and the total amount of traffic is as if the travellers respond to the marginal social costs (compare relationship (2.1.20)).

To obtain the optimal trip-matrix and the optimal assignment we can again use a stepwise assignment combined with a continuous adjustment of the trip-matrix. So we start with computing the least value of the marginal objective function for each relation with no traffic flow on the links:

$$(F'^{ab})_{x_{ij}=0} \quad \text{for all } ab \in P \tag{5.6.4}$$

We use this value to compute a first estimation of the trip-matrix $\mathbf{X}^{(0)}$ with:

$$g^{ab}(x^{ab(0)}) = (F'^{ab})_{x_{ij}=0} \quad \text{for all } ab \in P \tag{5.6.5}$$

We assign a part α_1 of the trip-matrix $\mathbf{X}^{(0)}$ to the network according to the routes found in relationship (5.6.4). Every link gets then a traffic flow $x_{ij}^{\alpha_1}$. Next we compute again those routes with the lowest values for the marginal objective function:

$$(F'^{ab})x_{ij} = x_{ij}^{\alpha_1} \quad \text{for all } ab \in P \tag{5.6.6}$$

We use those new values for F'^{ab} to compute a second estimation of the trip-matrix $\mathbf{X}^{(1)}$:

$$g^{ab}(x^{ab(1)}) = (F'^{ab})_{x_{ij}=x_{ij}^{\alpha_1}} \quad \text{for all } ab \in P \tag{5.6.7}$$

We now assign a part α_2 of the new matrix $\mathbf{X}^{(1)}$ to the network or rather:

$$(\alpha_1 + \alpha_2)\mathbf{X}^{(1)} - \alpha_1\mathbf{X}^{(0)} \tag{5.6.8}$$

according to the routes found in relationship (5.6.6). The process is continued until it has converged that is until:

$$g^{ab}(x^{ab(n)}) = (F'^{ab})_{x_{ij}=x_{ij}^{\alpha_n}} = g^{ab}(x^{ab(n-1)}) \quad \text{for all } ab \in P \tag{5.6.9}$$

Below we will prove that the process does in fact converge. Due to the assignment process (5.6.8):

$$x_{ij}^{\alpha_n} \geqslant x_{ij}^{\alpha_{n-1}} \quad \text{for all } ij \in L, \text{for all } n \tag{5.6.10}$$

Due to relationship (5.6.10) and the convexity of the function $F_{ij}^{\min}$:

$$(F'^{ab})_{x_{ij}=x_{ij}^{\alpha_n}} \geqslant (F'^{ab})_{x_{ij}=x_{ij}^{\alpha_{n-1}}} \quad \text{for all } ab \in P, \text{for all } n \tag{5.6.11}$$

Due to relationship (5.6.11) and 'demand functions' as (5.6.7):

$$x^{ab(n)} \leqslant x^{ab(n-1)} \quad \text{for all } ab \in P, \text{for all } n \tag{5.6.12}$$

Because there exists no upper bound for the vector of traffic flows and only the lower bound $\mathbf{0}$ for the trip-matrix, the vector of traffic flows and the trip-matrix approach each other and will match at the end of the process. Oscillations may occur in this process when (some elements of) relationship (5.6.8) are negative; however, these oscillations cannot occur when the steps are sufficiently small.

In Section 3.2.2.2 we saw that it is generally not possible to state demand functions only as a function of the users' costs for the relation itself (as is done in relationship (2.1.4)). Moreover we cannot state the total users' benefits by integration of the inverse demand function (as is done in relationship 3.2.4). However, we can still use the optimization algorithm described before, using a stepwise assignment, computing the least marginal objective function and using it continuously as input for a transport model, as can be readily understood. In this case, however, we must again be careful. Due to substitution effects it may well be possible, that the demand in some relation increases when at the same time the users' costs increase.

In the case of non-convex cost functions we saw that the stepwise assignment did not always yield a good solution for the costs minimization. Nor, of course, does the method yield a good solution for the surplus maximization in that case. Moreover we encounter the phenomenon of there being more possible trip-matrices, all satisfying the conditions for the trip-matrix.

Here we have solved a surplus maximization for a normative case. In reality such a trip-matrix will not occur, unless extra measures are taken, such as introducing charges and so on. Theoretically a method for a descriptive case would seem to be preferable. For the problem of costs minimization the use of a normative assignment has some practical advantages over the use of a descriptive assignment, for in this way it becomes possible to manage a complex system of constraints (the network constraints) with ease. Furthermore the use of a normative assignment instead of a descriptive one does not seem to be too harmful. Using a normative transport production and transport distribution instead of descriptive one seems more serious however. Moreover the use of a normative model instead of a descriptive one does not have any practical advantages; the same model must be used, only the input changes. So it does not seem to be too favourable to use the method described before for the surplus maximization problem. It may well be possible to use a descriptive transport production and distribution model and a normative assignment, thus getting an optimization on two levels, using the descriptive transport production and distribution for the benefit side of the objective function and the normative assignment for the costs side of the objective function. This needs to be further researched, however.

5.7 SOME EXPERIMENTS ON A SMALL NETWORK

5.7.1 Introduction

In the preceding sections we have seen that the use of the method of the stepwise assignment according to the least marginal objective function does not guarantee that the optimal solution is found. Moreover the solution found is dependent on the different assignment parameters, e.g. the number of steps, the magnitude of the parts assigned to the network and the size of the difference Δx. To get an idea of the operation of the method we will present now the results of some experiments on a small testing network. However, one must be very careful in drawing general conclusions from these experiments; for the results are dependent on several factors, such as:

the shape of the objective function;
the network;
the trip-matrices.

Moreover the accuracy required is dependent on the aims of the transport study and the position of the network optimization in the whole transport or land-use study.

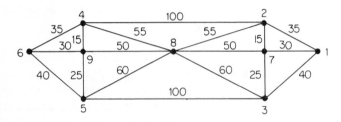

Figure 5.7.1   Network for the experiments

The network used consists of 9 nodes and 36 links and is shown in Figure 5.7.1. The geographical lengths of the (symmetric) links are also indicated in Figure 5.7.1. Four nodes, 2, 3, 4 and 5 are origins and destinations. The trip-matrix (called trip-matrix P4) is shown in Figure 5.7.2. The numbers of

	2	3	4	5
2	0	2,000	2,000	1,000
3	200	0	1,000	2,000
4	200	100	0	1,000
5	100	200	100	0

Figure 5.7.2   Trip-matrix P4 for the experiments

trips are chosen in such a way that the number in the return direction is always one tenth of the number of the outward direction, in order that the influence of the size of the numbers of trips may be studied. In Section 5.7.4

we will also use a trip-matrix P6 consisting of the trip-matrix of Figure 5.7.2 and two dominating relations 16 and 61.

First we will compare the results of the stepwise assignment according to the least marginal objective function with an infinite number of steps (400 actually) with the real optimal solution for a convex quadratic objective function (Section 5.7.2). Secondly we will research, for the same objective function, the sensitivity of the results for the different assignment parameters (Section 5.7.3). Finally we will examine the operation of the method for non-convex functions (Section 5.7.4).

5.7.2 Evaluation of the Solution for a Convex Quadratic Objective Function

We state only a master problem, for which the objective function is convex quadratic and always positive:

$$F = \sum_{ij \in L} \{\alpha_{ij} x_{ij} + \beta_{ij} x_{ij})^2\} \tag{5.7.1}$$

with:

$\alpha_{ij} = l_{ij} . 10^{-2}$ for all $ij \in L$ except $ij = 24$ and $ij = 42$

$\alpha_{ij} = 0$ for $ij = 24$ and $ij = 42$

$\beta_{ij} = l_{ij} . 10^{-5}$ for all $ij \in L$

The coefficients are chosen in such a way that the chances that the effect mentioned in Section 5.4.2 will occur, are high. The use of the links 24 and 42 is very profitable when there is little traffic. So there is a good chance that at the beginning of the assignment process almost all relations use the links 24 or 42. When the flows on these links become heavier, their use is no longer profitable; but a flow assigned in the first steps cannot now be re-assigned. So there is a fairly good chance that the stepwise assignment will not find the correct solution, even with the use of an infinite number of steps.

For the stepwise assignment 400 steps are used, all being the same size, namely 0·0025. For the marginal objective function difference quotients are used whose difference Δx equals 2·5.

To test whether the solution found is an optimal one, the same problem is stated and solved as a quadratic programming problem. To solve the quadratic programming problem we used a computer program developed at the Netherlands School of Economics in Rotterdam based on the algorithm of van de Panne and Whinston (1969).

For a straightforward formulation of the problem such as was given in the relationships (5.2.2), using the relationships (1.2.11) and (1.2.12) for the network constraints, 432 decision variables should be necessary, 108 equality constraints and 432 non-negativity constraints. To reduce the size of the program we considered the flows on the different paths as decision variables

Table 5.7.1 Paths used and numbers of trips on them for the solution found with quadratic programming and with stepwise assignment according to the least marginal objective function

Relation	Quadratic programming		Stepwise assignment according to the least marginal objective function	
	Path used	Number of trips on the path	Path used	Number of trips on the path
23	213	562	213	585[a]
	273	1438	273	1415
24	24	1694	24	1530
	2894	306	2894	115– 40[b]
			284	280–355
			2784	75– 0
			27894	0– 75
25	285	856	285	730–583
	2785	144	2785	0–147
			2895	50–197
			27895	147– 0
			2495	72
32	312	39	312	56
	372	161	372	144
34	384	918	384	582–257
	3894	82	3894	13–337
			3784	0–325
			37894	325– 0
			3724	80
35	35	1430	35	1420
	385	0	385	405–315
	3895	11	3895	0– 90
	3785	206	3785	60–150
	37895	353	37895	90– 0
			372495	25
42	42	200	42	200
43	483	100	483	68
			4273	28
			4213	4
45	465	275	465	292
	495	725	495	707
52	582	64	582	25
	5942	36	5942	75
53	53	199	53	173
	583	1	583	14
			594273	12
54	594	100	594	100

[a] All flows are rounded on one unity.
[b] Here the effect of Figure 5.7.3 occurs, with $x^{2894} = 115 - \delta$, $x^{284} = 280 + \delta$, $x^{2784} = 75 - \delta$, $x^{27894} = \delta$ and $0 \leqslant x \leqslant \delta$.

and we also excluded very unlikely paths in advance. In this way 66 decision variables are necessary, only 12 equality constraints and of course 66 non-negativity constraints.

First we give the values for the objective function:

F (stepwise assignment) = 17,160
F (quadratic programming) = 16,970

So two important facts can be seen:

(a) the stepwise assignment according to the least marginal objective function did not give the optimal solution;
(b) the difference from the optimal solution was only about 1 per cent.

To give a better analysis of the results it is necessary to look at the values for the decision variables, in this case the paths used and the traffic flows on them, and further to look at the resulting link flows. These are given in Table 5.7.1 and Table 5.7.2 respectively.

Table 5.7.2 Flows on the links computed with quadratic programming and with stepwise assignment according to the least marginal objective function

Link	Flows		Link	Flows	
	Quadratic programming	Stepwise assignment		Quadratic programming	Stepwise assignment
12	39	56	64	—	—
13	562	589	65	275	292
17	—	—	69	—	—
21	562	589	71	—	—
24	1,694	1,707	72	161	249
27	1,582	1,677	73	1,438	1,455
28	1,162	1,175	78	703	697
31	39	56	82	64	25
35	1,430	1,420	83	101	83
37	720	724	84	918	937
38	1,011	1,000	85	1,206	1,195
			87	—	—
42	236	319	89	752	740
46	275	292			
48	100	68	94	524	639
49	725	805	95	1,089	1,092
			96	—	—
53	199	173	98	—	—
56	—	—			
58	65	39			
59	136	187			

Before looking at these results we must note that the paths used are not always uniquely defined. For instance when some paths first fan out, come together then and fan out for a second time the flows on these paths are not

$X^{384} = \alpha$	$X^{384} = \alpha - \delta$
$X^{3784} = \beta$	$X^{3784} = \beta + \delta$
$X^{37894} = \gamma$	$X^{37894} = \gamma - \delta$
$X^{3894} = \alpha + \beta - \gamma$	$X^{3894} = \alpha + \beta - \gamma + \delta$

Figure 5.7.3   Different path flows (on the network of Figure 5.7.1) with the same value for the objective function ($\alpha, \beta, \gamma, \delta \geqslant 0, \gamma \leqslant \alpha + \beta, \delta \leqslant \gamma$)

uniquely defined. So the flows in Figure 5.7.3 give always exactly the same value for the objective function. The output of the program for the quadratic programming problem however gives only one single solution. The flows on the links are, in fact, uniquely defined though.

Looking at the two solutions we see that for all relations, except the relations 42 and 54, the paths used and the numbers of trips made on them are different. But the differences are not very large, especially not for the large flows. Moreover the quadratic programming gives only one solution.

The most important thing is that (almost) all differences between the solutions found with quadratic programming and those found with the stepwise assignment according to the least marginal objective function can be explained by the effect mentioned in Section 5.4.2 caused by the shape of the objective function for the links 24 and 42. In the beginning of the stepwise assignment it is very profitable to use these links. So we see that the relations 24, 25, 34 and even 35 in part use link 24 and the relations 42, 43, 52 and 53 use link 42 in the stepwise assignment. In the quadratic programming only relation 24 uses link 24 and the relations 42, 43 and 52 use 42. At the end of the stepwise assignment the flow on link 24 will become so large that it is also better to use other paths for relation 24. We see that even the path 27894 is used then. Of course it would have been better if this were not to happen. In the quadratic programming the number of trips of relation 24 in link 24 is about 10 per cent higher. In Figure 5.7.4 this effect is illustrated, taking the relevant paths of the relations 24 and 35.

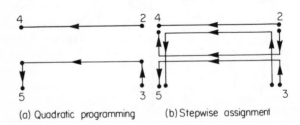

(a) Quadratic programming   (b) Stepwise assignment

Figure 5.7.4   Solutions found with quadratic programming and with stepwise assignment according to the least marginal objective function

It is interesting to see that the difference in the flow on link 24 is less than 1 per cent. The fact that more relations use the links 24 and 42 causes changes in the routes for other relations too. So the outward routes 273, 372 and 465 are more used in the stepwise assignment.

The differences in the flows on the links can be explained by the differences in the paths used. As said before the difference on the 'crucial' link 24 is less than 1 per cent. Among the heavy flows the largest differences are on the links 27, 49 and 94, respectively 6 per cent, 10 per cent and about 20 per cent. These differences are caused by different effects all related to the shape of the objective function on the links 24 and 42. The differences on all other well-used links are less than 1 per cent, and somewhat larger on the links with small flows. Generally we see the effect that 'errors made' compensate each other, especially on the links with an important effect on the objective function. Of course this is a very handy feature of the method used.

In conclusion we may state that the experiment shows that the stepwise assignment according to the least marginal objective function does not give the optimal solution, even with a convex objective function and the application of an infinite number of steps, but the differences between the solution found and the optimal one are very small, both for the objective function and for the decision variables. Moreover these differences can be explained and thus more or less predicted. As a last remark we must say that the necessary computation times are about the same for both the methods. This means that the stepwise assignment is—for this network—about hundred times faster when 4 instead of 400 steps are applied. Moreover the statement of a quadratic programming problem was only possible because of the very small size of the network.

5.7.3 Sensitivity for the Assignment Parameters

As stated in Section 5.4.3 there are three groups of assignment parameters for the stepwise assignment according to the least marginal objective function:

(a) the number of steps;
(b) the magnitude of the parts assigned to the network at the different steps (α_n);
(c) the magnitude of the difference Δx^n of the difference quotient.

The number of steps has a strong influence on the necessary computation time and as few as possible must therefore be taken. The other two sets of parameters must be chosen in coherence with the number of steps and must ensure that the solution found is as good as can be achieved with the number of steps used.

We varied all three sets of parameters for the problem (5.7.1) to get an impression of the sensitivity of the solution. We increased the number of steps from 1 to 400. The case of one step is special in that no account is taken of the interdependency of the flows of the different transport relations. One may look at the case of one step as a minimization of the objective function for every relation, while an estimate for the traffic flows is made in advance. This estimate equals Δx.

For a given number of steps we did four assignments:

(a)   equal steps, large value for Δx;
(b)   diminishing steps, large value for Δx;
(c)   equal steps, small value for Δx;
(d)   diminishing steps, small value for Δx.

The values of the parameters actually used are given in Table 5.7.3.

The values for the objective function computed with the different assignment parameters are given in Table 5.7.4. As expected we see a convergence to better solutions with an increasing number of steps. However, there are some oscillations; see that for a large Δx, the use of 10 or 15 steps gives a worse solution than the use of 7 steps for both equal and diminishing steps. For equal steps with large Δx the difference in the value of the objective function between 2 and 7 steps is about 10 per cent, between 7 and 400 steps is about 0·5 per cent and between 100 and 400 steps about 0·1 per cent.

Table 5.7.3   Values for the assignment parameters for the experiments

Number of steps	Large value of Δx	Small value of Δx	α_n, equal steps	α_n, diminishing steps
1	500, 1000, 2000	100	1×1	1
2	1000	100	2×0.5	0·7–0·3
3	750	75	0·4–0·3–0·3	0·6–0·3–0·1
4	500	50	4×0.25	0·5–0·3–0·15–0·05
5	400	40	5×0.2	0·4–0·2–0·2–0·1–0·1
7	300	30	2×0.15–5×0.14	0·3–0·2–0·16–0·13–0·1–0·07–0·04
10	200	20	10×0.1	0·2–0·18–0·16–0·14–0·11–0·08–0·06 0·04–0·02–0·01
15	150	15	10×0.7–5×0.6	0·15–0·13–0·12–0·1–0·09–0·08–0·07 0·06–0·05–0·04–0·03–0·03–0·02– 0·02–0·01
20	100	10	20×0.05	0·1–0·09–0·08–0·08–0·07–0·07–0·06 0·05–0·05–0·05–0·04–0·04–0·04 0·03–0·03–0·03–0·03–0·02–0·02 0·02
50	20		50×0.02	
100	10		100×0.01	
200	5		200×0.005	
300	4		$300 \times 1/300$	
400	2·5		400×0.0025	

Table 5.7.4   Values for the objective function computed with different assignment parameters

Number of steps	Large-value for Δx		Small value for Δx	
	Equal steps	Diminishing steps	Equal steps	Diminishing steps
1	28,556	28,556	52,996	52,996
2	19,343	20,132	23,739	31,376
3	19,791	18,791	20,894	26,488
4	17,676	18,231	19,042	22,869
5	17,660	18,280	18,486	20,487
7	17,254	17,660	18,032	19,105
10	17,585	18,302	17,600	18,308
15	17,376	17,818	17,369	17,934
20	17,244	17,516	17,319	17,516
50	17,193			
100	17,184			
200	17,167			
300	17,164			
400	17,160			

The sensitivity for the other assignment parameters decreases with an increasing number of steps. However the differences within the same number of steps are rather large. For 3 diminishing steps the value of the objective function with a small Δx is about 40 per cent higher than with a large Δx. For 10 steps the difference in the objective function between a small and a large value for Δx is still about 4 per cent.

Looking at the ultimate solution it is seen that the traffic flow on a link never exceeds 2,000, is greater than or equal to 1,000 only for 9 links and is quite often about 200. So with equal steps, the large value for Δx is a little larger than the largest flows assigned and the small value for Δx is of the same order as the majority of the flows assigned. For the diminishing steps the first parts assigned are a little larger than for the equal steps and the last parts assigned are smaller, so it can easily be seen how the Δx is related to the flows assigned.

We now see generally that the large values for Δx give better results than the small values and the equal steps give better results than the diminishing steps. In general the reason for this is that an error made in a large flow assigned is more serious than an error made in a small flow assigned. So the best results are obtained if Δx equals the large flows assigned, anyway for this type of objective function. There is a similar reason for the fact that the equal steps give a better result than the diminishing steps. With the diminishing steps a larger part is assigned in the first step, so a larger error can be made, which is more serious the larger the flows. With 15 or more steps the sensitivity for the assignment parameters is not so large, but still larger than that for the number of steps. Considering the fact that the number of steps is strongly connected with the computation time and the other assignment

Table 5.7.5(a)   Flows computed with different assignment parameters

Link	Optimal solution	1 step Δx large	1 step Δx small	2 steps Equal Δx large	2 steps Equal Δx small	2 steps Diminishing Δx large	2 steps Diminishing Δx small	3 steps Equal Δx large	3 steps Equal Δx small	3 steps Diminishing Δx large	3 steps Diminishing Δx small	4 steps Equal Δx large	4 steps Equal Δx small	4 steps Diminishing Δx large	4 steps Diminishing Δx small	5 steps Equal Δx large	5 steps Equal Δx small	5 steps Diminishing Δx large	5 steps Diminishing Δx small
12	39	—	—	—	100	—	60	—	120	20	80	—	150	—	100	—	120	40	120
13	562	—	—	1000	1000	600	600	600	600	600	600	500	525	600	700	400	820	600	600
17	—	—	—	—	—	—	—	—	—	—	—	—	—	—	—	—	—	—	—
21	562	4000	6000	1000	1000	600	600	600	600	600	600	500	525	600	700	400	820	600	600
24	1694	2100	2300	2000	3000	2800	4200	1600	2400	2400	3600	1500	1500	2000	3000	1600	1600	1600	2400
27	1582	—	—	1050	1150	1470	1610	2340	2420	1760	1880	2300	2325	1900	1900	1840	1860	2040	2120
28	1162	—	—	1500	1500	900	900	900	900	900	900	1000	1500	1050	1050	1400	1400	1200	1200
31	39	2000	3200	—	100	—	60	—	120	20	80	—	150	40	100	—	120	40	120
35	1430	1200	—	1000	1000	1400	600	800	600	1200	600	1500	1250	1100	700	1200	1400	1000	900
37	720	—	—	700	1600	900	2240	1500	1280	1080	1920	700	800	1110	1600	1200	680	1160	1280
38	1011	—	—	1500	500	900	300	900	1200	900	600	1000	1000	950	800	800	1000	1000	900
42	236	400	600	350	400	340	480	310	390	320	440	325	375	330	400	320	340	320	380
46	275	—	—	500	500	300	300	300	600	300	400	250	500	350	500	200	400	400	600
48	100	—	—	50	50	30	30	60	60	40	40	50	50	50	50	60	60	60	60
49	725	2000	4000	1000	2000	1400	2800	1100	1600	1300	2400	1000	1250	1150	2000	1000	1200	1000	1600
53	200	200	—	200	100	200	60	200	120	200	80	200	150	190	100	200	160	180	120
56	—	—	—	—	—	—	—	—	—	—	—	—	275	—	—	—	20	—	10
58	65	—	—	—	50	—	30	30	30	40	40	25	25	30	50	20	40	40	40
59	136	200	400	200	250	170	310	170	250	160	280	175	200	180	250	180	380	180	330
64	—	—	—	—	—	—	—	—	—	—	—	—	—	—	—	—	20	—	10
65	275	—	—	500	500	300	300	300	600	300	400	250	500	350	500	200	400	400	600
69	—	—	—	—	—	—	—	—	—	—	—	—	—	—	—	—	—	—	—
71	161	1200	3200	700	1600	900	2240	600	1280	780	1920	450	800	660	1600	400	680	560	1280
72	1438	3100	2300	1050	1150	1470	1610	1440	1520	1460	1580	1550	1575	1450	1450	1640	1260	1440	1520
73	703	—	—	—	—	—	—	1800	900	600	300	1000	750	900	450	1000	600	1200	600
78	—	—	—	—	—	—	—	—	—	—	—	—	—	—	—	—	—	—	—
82	64	—	—	—	50	—	30	30	30	40	40	25	25	20	50	20	40	20	40
83	101	—	—	50	50	30	30	60	60	40	40	50	50	60	50	60	60	80	60
84	918	—	—	1500	1500	900	900	900	900	900	900	1000	1250	1050	900	1200	1200	1200	1100
85	1206	—	—	1500	500	900	300	900	1200	900	600	1000	1250	950	800	1400	1200	1000	1000
87	—	—	—	—	—	—	—	—	—	—	—	—	—	—	—	—	—	—	—
89	752	200	400	200	250	170	310	1800	900	600	300	1000	750	900	600	600	600	1200	600
94	524	2000	4000	200	200	170	310	1070	1150	460	580	925	950	630	850	580	980	780	930
95	1089	—	—	1000	2000	1400	2800	2000	1600	1600	2400	1250	1250	1600	2000	1200	1200	1600	1600
96	—	—	—	—	—	—	—	—	—	—	—	—	—	—	—	—	—	—	—
98	—	—	—	—	—	—	—	—	—	—	—	—	—	—	—	—	—	—	—
objective	16,970	28,556	52,996	19,343	23,739	20,132	31,376	19,791	20,894	18,791	26,488	17,676	19,042	18,231	22,869	17,660	18,486	18,280	20,487

Link	Optimal solution	7 steps Equal Δx large	7 steps Equal Δx small	7 steps Dim. Δx large	7 steps Dim. Δx small	10 steps Equal Δx large	10 steps Equal Δx small	10 steps Dim. Δx large	10 steps Dim. Δx small	15 steps Equal Δx large	15 steps Equal Δx small	15 steps Dim. Δx large	15 steps Dim. Δx small	20 steps Equal Δx large	20 steps Equal Δx small	20 steps Dim. Δx large	20 steps Dim. Δx small
12	39	56	86	68	140	40	60	104	132	38	62	82	100	30	60	62	80
13	562	560	574	600	700	610	610	620	638	547	667	612	632	605	605	588	608
17	—	—	—	—	—	—	—	—	—	—	—	—	—	—	—	—	—
21	562	560	574	600	700	610	610	620	638	547	667	612	632	605	605	588	608
24	1694	1720	1740	1680	1880	1600	1600	1760	1760	1740	1740	1740	1740	1700	1650	1720	1740
27	1582	1750	2060	1910	2100	1950	1850	2048	2030	1892	1652	1948	1928	1740	1750	1857	1827
28	1162	1150	1150	1140	1330	1200	1300	1250	1250	1080	1200	1220	1220	1150	1250	1190	1180
31	39	56	86	68	140	40	60	104	132	38	62	82	100	30	60	62	80
35	1430	1420	1280	1280	1210	1400	1400	1380	1340	1340	1340	1400	1380	1400	1400	1420	1400
37	720	854	704	912	960	860	740	736	788	772	798	728	730	720	740	728	730
38	1011	870	1130	940	890	900	1000	980	940	1050	1000	990	990	1050	1000	990	990
42	236	302	332	296	375	330	330	332	346	319	325	329	329	315	325	320	325
46	275	280	430	370	530	300	300	380	410	320	330	330	370	300	300	310	320
48	100	70	56	70	50	60	60	62	62	65	65	60	60	65	65	65	65
49	725	870	1020	930	1370	1000	1000	1220	1190	890	880	1120	1080	850	850	990	980
53	200	200	170	192	140	180	160	160	158	174	174	170	164	180	170	174	168
56	—	—	28	—	21	—	—	15	32	—	—	—	20	—	—	—	—
58	65	28	42	42	34	30	50	46	34	42	36	41	47	40	40	41	42
59	136	172	300	166	335	190	190	319	316	184	190	189	269	180	190	185	190
64	—	—	—	—	—	—	—	—	—	—	—	—	—	—	—	—	—
65	275	280	430	370	530	300	300	380	410	320	330	330	370	300	300	310	320
69	—	—	—	—	—	—	—	—	—	—	—	—	—	—	—	—	—
71	161	294	564	432	960	460	440	696	668	372	348	568	550	320	340	438	420
72	1438	1470	1500	1430	1410	1450	1450	1458	1440	1502	1382	1458	1438	1440	1450	1467	1447
73	703	840	700	960	690	900	700	630	710	790	720	650	670	700	700	680	690
78	—	—	—	—	—	—	—	—	—	—	—	—	—	—	—	—	—
82	64	28	42	34	34	30	30	46	32	30	24	41	41	30	30	35	30
83	101	70	56	78	50	60	80	62	64	77	77	60	66	75	75	71	77
84	918	1010	1010	1140	1110	1000	900	1020	1020	940	880	1000	970	950	900	970	960
85	1206	1150	1270	940	1020	1200	1100	1160	1180	1200	1260	1150	1200	1200	1200	1170	1160
87	—	—	—	—	—	—	—	—	—	—	—	—	—	—	—	—	—
89	752	700	700	960	780	800	1000	680	700	780	780	710	710	750	850	720	740
94	524	592	860	646	1115	890	990	999	996	714	780	899	909	680	790	795	790
95	1089	1150	1160	1410	1370	1100	1200	1220	1210	1140	1070	1120	1150	1100	1100	1100	1120
96	—	—	—	—	—	—	—	—	—	—	—	—	—	—	—	—	—
98	—	—	—	—	—	—	—	—	—	—	—	—	—	—	—	—	—
objective function	16,970	17,254	18,032	17,660	19,105	17,585	17,600	18,302	18,308	17,376	17,369	17,818	17,934	17,244	17,319	17,516	17,516

parameters are not, it is obvious that a proper choice of these two other sets
of parameters is very important.

Table 5.7.6   Flows computed with different numbers of steps

Link	Optimal solution	20	50	100	200	300	400
12	39	30	52	56	56	55	56
13	562	605	604	584	584	584	589
17	—	—	—	—	—	—	—
21	562	605	604	584	584	584	589
24	1694	1700	1720	1710	1710	1713	1707
27	1582	1740	1660	1690	1685	1676	1677
28	1162	1150	1180	1180	1175	1177	1175
31	39	30	52	56	56	55	56
35	1430	1400	1400	1400	1420	1413	1420
37	720	720	768	734	724	725	724
38	1011	1050	980	1010	1000	1007	1000
42	236	315	318	318	318	319	319
46	275	300	300	290	290	293	292
48	100	65	68	68	68	68	68
49	725	850	820	820	810	807	805
53	199	180	172	174	173	173	173
56	—	—	—	—	—	—	—
58	65	40	42	40	40	40	39
59	136	180	186	186	186	187	187
64	—	—	—	—	—	—	—
65	275	300	300	290	290	293	292
69	—	—	—	—	—	—	—
71	—	—	—	—	—	—	—
72	161	320	268	264	254	251	249
73	1438	1440	1440	1460	1460	1460	1455
78	703	700	720	700	695	690	697
82	64	30	26	26	25	25	25
83	101	75	84	82	83	83	83
84	918	950	940	940	930	933	937
85	1206	1200	1200	1210	1195	1200	1195
87	—	—	—	—	—	—	—
89	752	750	740	740	745	740	740
94	524	680	646	646	646	640	639
95	1089	1100	1100	1100	1095	1093	1092
96	—	—	—	—	—	—	—
98	—	—	—	—	—	—	—
Objective function	16,970	17,244	17,193	17,184	17,167	17,164	17,160

For a deeper analysis of the sensitivity for the assignment parameter
we must look at the values for the decision variables: the flows on the links
These flows are given in the Tables 5.7.5 and 5.7.6.

With one step the use of the values 500, 1,000 or 2,000 for Δx, all give the same solution, in which all relevant relations except 35 and 53 use the links 24 or 42. When the small value 100 for Δx is used, all relevant relations use the links 24 or 42, including the relations 35 and 53. It is obvious that the application of one step does not give a good solution. The flow on the 'crucial' link 24 is more than two or three times too high, while important links like 84, 85 and 89 do not get any traffic flow. So obviously *stepwise* assignment is necessary.

Looking at the link flows for the different assignments with more steps, we see that again the use of the link 24 is critical. Moreover there are differences in the way the diagonals are used. Generally speaking the most important 'error' made is too much flow on link 24. The flows on the diagonals are fluctuating more or less around the optimal solution.

With two steps we see that with a large value for Δx the general picture of the optimal solution is already given, though with large margins. The application of small steps causes much too much flow on link 42. The application of three steps gives an improvement, especially in diminishing steps. Still the errors made are fairly large, including the ones on the critical link 24. The smaller flows are better defined now. With four steps the optimal solution is fairly well approached, at least when equal steps and a large value for Δx are applied. Only for the flow on link 27 is a rather large error made. Five steps give a better solution again, in which the flow on link 27 is better determined.

For large flows ($> 1,000$) the differences between the solutions found with 7 and with 400 steps are less than 5 per cent (equal steps, large value for Δx). The sensitivity for the other assignment parameters is still rather high.

The application of 10, 15 and 20 steps does not give a significant improvement in the equal steps and with a large value for Δx. But the sensitivity for the assignment parameters decreases strongly. Looking at the objective function we see that the difference from the optimal solution decreases from 8 per cent via 6 per cent to 3 per cent in diminishing steps and with a small value for Δx. The same differences for the flow on link 27 are respectively 28 per cent, 22 per cent and 15 per cent. When more than 20 steps are used (50, 100, 200, 300, 400) the solutions converge further, the differences becoming smaller and smaller but not vanishing.

In conclusion we may state that a proper choice of the assignment parameters is very important. Especially important is the choice of the size of the difference Δx^n and of the parts α^n to be assigned to the network. If these two sets of parameters are chosen in the right way, it seems to be possible to use a fairly small number of steps. So the use of four steps seems to lead to quite reasonable results for the network, trip-matrix and objective function for the experiment.

5.7.4 The Operation of the Method for Non-convex Functions

When the objective function is not completely convex there are a few additional reasons why the application of the stepwise assignment according to the least marginal objective function does not guarantee an optimal solution. The condition that the marginal objective function on all paths used is to be equal and smaller than on all paths not used is not necessary or sufficient for a minimum solution for a non-convex objective function. In the case of convex functions the optimal solution is generally reached when the flows are divided among several paths, while in the case of concave functions a concentration of different relations on the same links is generally profitable.

Still we suggested in Section 5.4.2 that the stepwise assignment according to the least marginal objective function will also give reasonable solutions in the case of non-convex functions. In this subsection we will research this statement further for a special case.

Once more we will work with the network of Figure 5.7.1 and the trip-matrix P4 of Figure 5.7.2. In addition we will use the trip-matrix of Figure 5.7.5, called matrix P6, which consists of matrix P4 augmented with two dominating relations 16 and 61.

	1	2	3	4	5	6
1	0	0	0	0	0	10,000
2	0	0	2,000	2.000	1,000	0
3	0	200	0	1,000	2,000	0
4	0	200	100	0	1,000	0
5	0	100	200	100	0	0
	1,000	0	0	0	0	0

Figure 5.7.5   Trip-matrix P6 for the experiments

We will now state the complete problem (problem 5.2.1)) with the dimensions and the flows as decision variables:

$$\min_{c_{ij}, x_{ij}} \sum_{ij \in L} F_{ij}(x_{ij}, c_{ij}) \tag{5.2.1}$$

subject to $A\mathbf{X} = \mathbf{0}, \mathbf{X} \geqslant \mathbf{0}, \mathbf{C} \geqslant \mathbf{C}^{\min}$

in which the objective function for a link has the following form:

$$F_{ij} = x_{ij}t_{ij}(x_{ij}, c_{ij}) + i_{ij}(c_{ij}) \tag{5.7.2}$$

with:

$$t_{ij} = l_{ij}\left\{ \text{TZERO} + \text{CCR}\left(\frac{x_{ij}}{c_{ij}}\right)^2 \right\}$$

$$i_{ij} = l_{ij}\alpha_{ij}(c_{ij} - c_{ij}^{\min})^{1/2}$$

with the coefficients TZERO, CCR and α_{ij}

Four different solution methods are applied to this problem (Steenbrink, 1970):

(a) stepwise assignment according to the least marginal objective function;
(b) continuous optimal adjustment (see Section 4.4.2.1);
(c) stepwise assignment according to the least marginal objective function followed by a descriptive assignment and an optimal adjustment;
(d) manual choice of some most probable optimal network structures followed by a descriptive assignment and an optimal adjustment and selection of the best solution (a more or less interactive system, see Section 4.4.4)

For the stepwise assignment according to the least marginal objective function four different sets of assignment parameters are used; see Table 5.7.7.

Table 5.7.7   Values for the assigned parameters for the stepwise assignment according to the least marginal objective function

Number of steps	α_n	Δx
3	0·6–0·3–0·1	500
3	0·6–0·3–0·1	100
4	4×0.25	500
4	4×0.25	100

The mere continuous optimal adjustment is started with a network in which all links have a very large dimension. Finally the descriptive assignment is in accordance with the second principle of Wardrop (individual users' costs minimization, see Section 2.1.3) and is computed with a capacity-restraint process using four steps (0·1–0·4–0·3–0·2). (The first step of this process is chosen small, because some links may have a very small dimension.)

Decomposition of the problem gives the n_L subproblems:

$$\left.\begin{aligned} F_{ij}^{\min}(x_{ij}) = \min_{c_{ij}} x_{ij}t_{ij}(x_{ij}, c_{ij}) + i_{ij}(c_{ij}) \\ \text{subject to } c_{ij} \geqslant c_{ij}^{\min} \end{aligned}\right\} \text{ for all } ij \in L \qquad (5.7.3)$$

and the master problem:

$$\min_{x_{ij}} \sum_{ij \in L} F_{ij}(x_{ij}) \text{ subject to } AX = 0, \; X \geqslant 0 \qquad (5.7.4)$$

Solution of the subproblems (5.7.3) gives:

$$
F_{ij}^{\min}(x_{ij}) = \begin{cases}
= l_{ij}x_{ij}\left\{\text{TZERO} + \text{CCR}\left(\dfrac{x_{ij}}{c_{ij}^{\min}}\right)^2\right\} \\[4pt]
\qquad \text{if } x_{ij} \leqslant (c_{ij}^{\min})^{5/6}\left(\dfrac{\alpha_{ij}}{4\,\text{CCR}}\right)^{1/3} \\[16pt]
= l_{ij}\,\text{TZERO}\,x_{ij} + l_{ij}(4^{1/5} + 4^{-4/5})\text{CCR}^{1/5}\alpha_{ij}^{4/5}x_{ij}^{3/5} - \alpha_{ij}(c_{ij}^{\min})^{1/2} \\[4pt]
\qquad \text{if } x_{ij} > (c_{ij}^{\min})^{5/6}\left(\dfrac{\alpha_{ij}}{4\,\text{CCR}}\right)^{1/3}
\end{cases}
\tag{5.7.5}
$$

So if $c_{ij}^{\min} = 0$ the objective function for the master problem on a link is concave and if $c_{ij}^{\min} > 0$ the function is first convex and then concave.

Three situations are examined now:

(a)   all links same characteristics;
(b)   one existing road;
(c)   one cheap road.

Situation (a): all links same characteristics

$$(\alpha_{ij} = \alpha, \ c_{ij}^{\min} = 0 \quad \text{for all } ij \in L)$$

Trip-matrix P4 (Table 5.7.8)

The interactively-found best solution is the construction and use of the rectangle 2453 (i.e. the links 24, 42, 45, 54, 53, 35, 32 and 23). All three methods used give almost the same solution, in which the rectangle and the diagonals are constructed and used.

Table 5.7.8   Values for the objective function for the situation with all links same characteristics and trip-matrix P4

Assignment parameters	SALMOF[a]	SALMOF–DESCASS–OPTADJ[a]
3 steps, $\Delta x = 500$	2,805	2,805
3 steps, $\Delta x = 100$	2,799	2,799
4 steps, $\Delta x = 500$	2,805	2,805
4 steps, $\Delta x = 100$	2,790	2,796
Continuous optimal adjustment		2,805
Only roads 273, 495 and 987		2,822
Best solution found		2,620

[a] SALMOF: stepwise assignment according to the least marginal objective function.
DESCASS: descriptive assignment.
OPTADJ: optimal adjustment.

Trip-matrix P6 (Table 5.7.9)

The interactively-found best solution is the construction of the main road 17896 (i.e. the links 17, 71, 78, 87, 89, 98, 96 and 69) and the roads 495 and 273. All applications of SALMOF result in the construction and use of all links, except 213 and 465, while in the optimal adjustment the route 1246 and vice versa is constructed and used.

Table 5.7.9   Values for the objective function for the situation with all links same characteristics and trip-matrix P6

Assignment parameters	SALMOF	SALMOF–DESCASS–OPTADJ
3 steps, $\Delta x = 500$	6,996	6,997
3 steps, $\Delta x = 100$	6,986	7,001
4 steps, $\Delta x = 500$	6,951	6,974
4 steps, $\Delta x = 100$	6,921	6,983
Continuous optimal adjustment		7,246
Without 213, 465 and diagonals		6,772
Best solution found		6,633

Situation (b): one existing road

The road 24 (links 24 and 42) is considered as an existing road with $c_{24}^{\min} = c_{42}^{\min} = 2,000$; furthermore all $\alpha_{ij} = \alpha$ for all $ij \in L$ and $c_{ij}^{\min} = 0$ for all $ij \in L$ except $ij = 24$ and $ij = 42$.

Trip-matrix P4 (Table 5.7.10)

The interactively-found best solution is that all relations except 23, 32, 45, 54 and (parts of) 35 and 53 use the existing road. This solution is also found by SALMOF. Small differences occur in the amount of relation 53 that uses the existing road. With the continuous optimal adjustment the diagonals are also constructed and used.

Table 5.7.10   Values for the objective function for the situation of one existing road and trip-matrix P4

Assignment parameters	SALMOF	SALMOF–DESCASS–OPTADJ
3 steps, $\Delta x = 500$	2,278	2,277
3 steps, $\Delta x = 100$	2,278	2,277
4 steps, $\Delta x = 500$	2,278	2,277
4 steps, $\Delta x = 100$	2,264	2,277
Continuous optimal adjustment		2,468
Best solution found		2,264

Trip-matrix P6 (Table 5.7.11)

The interactively-found best solution is that the large relations x^{16} and x^{61} use the existing road totally; for the rest the solution is the same as for trip-matrix P4. This solution is also found by SALMOF. With the continuous optimal adjustment only one fifth of x^{16} uses the existing road and once more the diagonals are constructed and used.

Table 5.7.11   Values for the objective function for the situation of one existing road and trip-matrix P6

Assignment parameters	SALMOF	SALMOF–DESCASS–OPTADJ
3 steps, $\Delta x = 500$	6,301	6,302
3 steps, $\Delta x = 100$	6,301	6,302
4 steps, $\Delta x = 500$	6,301	6,302
4 steps, $\Delta x = 100$	6,294	6,302
Continuous optimal adjustment		6,815
Best solution found		6,294

Situation (c): one cheap road

The investment costs for road 24 (links 24 and 42) are half as high as for the other links:

$$\alpha_{24} = \alpha_{42} = \tfrac{1}{2}\alpha; \qquad \alpha_{ij} = \alpha \quad \text{for all } ij \in L \text{ except } ij = 24 \quad \text{and} \quad ij = 42;$$

$$c_{ij}^{\min} = 0 \quad \text{for all } ij \in L$$

Trip-matrix P4 (Table 5.7.12)

The interactively-found best solution is that only the rectangle 2453 is constructed and used, while x^{52}, x^{25}, x^{34} and x^{43} use the cheap road. This solution is also found by SALMOF. With the continuous optimal adjustment the diagonals are also constructed and used.

Table 5.7.12   Values for the objective function for the situation of one cheap road and trip-matrix P4

Assignment parameters	SALMOF	SALMOF–DESCASS–OPTADJ
3 steps, $\Delta x = 500$	2,740	2,739
3 steps, $\Delta x = 100$	2,740	2,739
4 steps, $\Delta x = 500$	2,740	2,739
4 steps, $\Delta x = 100$	2,740	2,739
Continuous optimal adjustment		3,058
Best solution found		2,739

Trip-matrix P6 (Table 5.7.13)

The interactively-found best solution is that the large relations x^{16} and x^{61} use the cheap road; further the solution is identical with that for trip-matrix P4. This solution is also found by SALMOF. With the continuous optimal adjustment only one third of the trips x^{16} use the cheap road, the other two thirds of x^{16} using route 17896. Moreover the diagonals are constructed and used when the continuous optimal adjustment is applied.

Table 5.7.13   Values for the objective function for the situation of one cheap road and trip-matrix P6

Assignment parameters	SALMOF	SALMOF–DESCASS–OPTADJ
3 steps, $\Delta x = 500$	6,940	6,941
3 steps, $\Delta x = 100$	6,940	6,941
4 steps, $\Delta x = 500$	6,940	6,941
4 steps, $\Delta x = 100$	6,940	6,941
Continuous optimal adjustment		8,095
Best solution found		6,940

Conclusions

Looking at the results we see that in four out of the six cases the best solution found is also found with the stepwise assignment according to the least marginal objective function. Only in the first situation, when all links have the same characteristics, does the stepwise assignment according to the least marginal objective function not give the best solution. But in that first situation the very use of the concavity of the functions is the only way to get a good solution. It is to be expected then that the correct solution is not found.

Still it is interesting to see that for *all* cases the results of the stepwise assignment according to the least marginal objective function are at least as good as those found with the heuristic method of the continuous optimal adjustment and mostly better. This is especially true for the cases with the dominating transport relations. The stepwise assignment according to the least marginal objective function followed by a descriptive assignment and an optimal adjustment generally gives higher values for the objective function than the use of the method on its own. This was to be expected.

Also in these experiments we see that the solutions are sensitive to the different assignment parameters. As expected, the application of four steps gives better results than the use of three steps. For Δx the small value generally gives better results than the large one, in connection with the concavity of the functions.

As a general conclusion we may state that it may be expected that the application of the stepwise assignment according to the least marginal objective function will generally give reasonable results, even in the case of a non-convex objective function.

REFERENCES

Allais, M., Del Viscovo, M., Duquesne de la Vinelle, L., Oort, C. J., and Seidenfus, H. St. (1965). *Options de la Politique Tarifaire dans les Transport*. Communauté Economique Européenne, Collection Etudes, Série Transports, No. 1, Brussels.

Beckmann, M. J., McGuire, C. B., and Winsten, C. B. (1956). *Studies in the Economics of Transportation*, Cowles Commission, Yale University Press, New Haven.

Bruynooghe, M. (1967). *Affectation du Trafic sur un Multi-réseau*, Institut de Recherche des Transport, Arcueil, November.

Bruynooghe, M., Gibert, A., and Sakarovitch, M. (1968). Une méthode d'affectation du trafic. Paper presented at the Fourth International Symposium on *The Theory of Traffic Flow*, Karlsruhe (June).

Dafermos, S. C., and Sparrow, F. T. (1969). The traffic assignment problem for a general network. *Journal of Research of the National Bureau of Standards: B. Mathematical Sciences*, **73B**, No. 2 (April–June).

Gilbert, A. (1968a). *A Method for the Traffic Assignment Problem when Demand is Elastic*, Transport Network Theory Unit Report LBS-TNT 85, London Business School, London.

Gilbert, A. (1968b). *A Method for the Traffic Assignment Problem*, Transport Network Theory Unit Report LBS-TNT 95, London Business School, London.

Murchland, J. D. (1969). *Gleichgewichtsverteilung des Verkehrs im Strassennetz*. Verlag Anton Hain Meisenheim. Also in English (1969): *Road Network Traffic Distribution in Equilibrium*. Paper presented at the conference: 'Mathematical Methods in the Economic Science', Oberwolfach.

Nederlands Economisch Instituut (1972). *Integrale Verkeers- en Vervoerstudie* (*annex V*). Staatsuitgeverij, 's-Gravenhage.

Oort, C. J. (1966). *De Infrastruktuur van het Vervoer*, Algemene Verladers—en Eigen Vervoerders Organisatie E.V.O., 's-Gravenhage.

Panne, C. van de, and Whinston, A. (1969). The symmetric formulation of the Simplex method for quadratic programming. *Econometrica*, **37**, No. 3 (July).

Steel, M. A. (1965). Capacity restraint, a new technique. *Traffic Engineering and Control* (October).

Steenbrink, P. A. (1970). *Optimalisering van de Infrastruktuur: Notitie 5. Stapsgewijze Oplading van het Netwerk volgens de op dat Moment 'Goedkoopste' Route: Eerste Uiteenzetting*. Internal note of the Nederlandse Spoorwegen, OP2/239/41 (333)G, Utrecht.

Steenbrink, P. A. (1971). Optimalisering van de infrastruktuur. *Verkeerstechniek*, **22**, No. 7 (July).

6

Extensions of the Problem

In this chapter we will discuss more extended transport network optimization problems. In the first two sections we will work with more trip-matrices: in Section 6.1 the dynamic situation is considered and in Section 6.2 networks for more travel modes. In the last section of this chapter some remarks are made about the optimal structure of a network.

6.1 OPTIMIZATION OVER TIME

6.1.1 Statement of the Problem

In the two preceding chapters we discussed the static problem in which *one* transport network had to be constructed for *one* situation. But the transport network optimization problem is essentially a dynamic problem. *Many* transport networks have to be made for *many* situations at different times and these networks and situations have strong relationships with each other. Stated simply the problem is as follows: the transport network can change only very slowly, while the factors that influence the transport and so the trip-matrices themselves change more rapidly. This means that in deciding what transport network to build or what changes to make to the existing network and when, we have to take into account the transport situation for many years.

The very long lifetime of the connections in the transport network has other effects. Because of the long lifetime it is possible to make very high investments that can be 'paid back' over a long period. These very high investments are necessary because of the high construction costs of the connections in the transport network.

Besides the financing and depreciation aspects of the long lifetime of infrastructure there are certain other aspects which are probably even more important. There is the fact that the construction or improvement of a transport connection has an influence on the whole system that lasts many years and is also, almost always, irreversible. In addition to the lasting effects on transport production and distribution, the choice of route and travel mode, there is the very important effect on the spatial structure and the (regional) economics.

Finally, we should mention some more dynamic aspects such as the fact that a road is more expensive to construct in stages than all at once, the fact that the construction of a road may take many years, and the fact that budgetary or other constraints can be imposed annually.

So it is clear that a dynamic problem has to be solved. In this section we will state the problem and mention some possible solution methods. We can state the problem as follows:

surplus maximization, descriptive case:

$$\max_{C(t)} \int_{t_0}^{t_e} S_t(\mathbf{X}(t), \mathbf{C}(t))\, dt$$

subject to:

$$\left. \begin{array}{l} A\mathbf{X}(t) = \mathbf{0}, \mathbf{X}(t) \geqslant \mathbf{0} \\[4pt] \mathbf{G}_t(\mathbf{X}(t), \mathbf{C}(t)) = \mathbf{0} \\[4pt] \mathbf{H}(\mathbf{X}(t), \mathbf{C}(t)) >, =, < \mathbf{0} \end{array} \right\} \quad \text{for } t_0 \leqslant t \leqslant t_e \qquad (6.1.1)$$

in which:

$\mathbf{C}(t)$—(vector of functions of the) dimensions of the links
$\mathbf{X}(t)$—(vector of functions of) traffic flows
S_t —social surplus function for period t
$\mathbf{G}_t$ —set of functions describing the behaviour of the travellers in period t, including the influence factors of t
$\mathbf{H}$ —set of remaining constraints
t_0 —starting point of time
t_e —ending point of time (can be infinity of course)

Thus formulated the problem has become the maximization of a functional, a problem treated in the calculus of variations (Pontryagin, 1961). (A functional assigns a number to every function or set of functions on which it depends.) When we assume that everything is constant for a certain time period and that the situation may change at the end of the period, remain constant again during the following time period and so on, we get the following formulation:

$$\max_{\mathbf{C}_1, \dots, \mathbf{C}_n} \sum_{t=1}^{n} S_t(\mathbf{X}_t, \mathbf{C}_t)$$

subject to:

$$AX_t = 0 . X_t \geqslant 0$$
$$G_t(X_t, C_t) = 0 \qquad \text{for all } t = 1, \ldots, n \qquad (6.1.2)$$
$$H(X_t, C_t) >, =, <0$$

This again is an optimization problem for a function of variables. However, the number of variables and constraints is much more than for the static problem.

Of course it is also possible to state the problems (4.1.2), (4.1.3) and (4.1.4) with a fixed trip-matrix and/or a normative assignment in a dynamic way. Looking at the problem as stated in relationship (6.1.2) we can try to split it into smaller problems. One possibility is splitting the dynamic problem into n static problems, one for each time period. We get then the comparatively static problems:

$$\max_{C_t} S_t(X_t, C_t)$$

subject to:

$$AX_t = 0, X_t \geqslant 0$$
$$G_t(X_t, C_t) = 0 \qquad \text{for all } t = 1, \ldots, n \qquad (6.1.3)$$
$$H_t(X_t, C_t) >, =, <0$$

Note that problem (6.1.3) looks very similar to problem (6.1.2). The main difference is in the third set of constraints which is H in problem (6.1.2) and n times H_t in problem (6.1.3). This, however, is also the essential set of constraints that makes the dynamic problem different from n comparatively static problems. All dynamic aspects mentioned at the beginning of this section are included in this set of constraints.

Another way to simplify the dynamic network optimization problem is to focus particularly on the dynamic aspects of time budgeting and ignore the interdependency of the various links in the network. Several authors use this approach. De Neufville (1969) and Mori (1968) for instance assume that all benefits and costs for all roads for all years have been defined in advance. They then use dynamic programming (see Section 6.1.2) to compute the optimal staging of the construction of the network subject to an annual budget constraint if desirable. But in this approach what is essential for network optimization, namely the interdependency of the links in the network, is ignored.

Before turning to the solution techniques we must say something more about the objective function. As formulated in problem (6.1.1.) or (6.1.2) only the surpluses from t_0 to t_e or from 1 to n are computed. But of course the time after t_e or n is also important. To be strictly correct, we have to take t_e

or n equal to infinity. But this would not only make the size of the problem prohibitive it would also be impossible to predict the influence factors for dates far into the future or to predict the future trip-matrices in the case of costs minimization. Fortunately the surpluses far into the future do not count so heavily in the objective function when the present value of the surpluses is used:

$$S_t = S_t' e^{-\pi t} \quad \text{or} \quad S_t = S_t' \left(\frac{1}{1 + \pi} \right)^t$$

in which:

S_t'—surplus in year t, undiscounted
π—discount rate

The discounting factors are continuously decreasing to their limit value 0:

$$\lim_{t \to \infty} e^{-\pi t} = 0 \quad \text{and} \quad \lim_{t \to \infty} \left(\frac{1}{1 + \pi} \right)^t = 0$$

So the further into the future a period is the less support that period gives to the objective function. This fact makes it possible to make an estimate of the value of the objective function for the period from t_e to infinity, if the network for t_e is $C(t_e)$. This value is called the scrap value of the network.

In defining this scrap value we must make some assumptions about the behaviour of the travellers and of the decision makers after the 'final' time period t_e. We might suppose that after this last time period the regulating authority will invest in new roads if the social surpluses of these investments are big enough. In that case the social surpluses will not change all that much from the last time period till infinity. Using this assumption (as did, for instance, Bergendahl, 1969a) we get the following value for the scrap value:

$$S_s = \int_{t_e}^{\infty} S_t \, e^{-\pi t} \, dt = \int_{t_e}^{\infty} S_{t_e} \, e^{-\pi t} \, dt = S_{t_e} \frac{e^{-\pi t_e}}{\pi}$$

or:

$$S_s = \sum_{t=n+1}^{\infty} S_t \frac{1}{(1 + \pi)^t} = \sum_{t=n+1}^{\infty} S_{t_e} \frac{1}{(1 + \pi)^t} = S_{t_e} \frac{1}{\pi(1 + \pi)^n} \approx S_{t_e} \frac{e^{-\pi t_e}}{\pi} \quad (6.1.4)$$

6.1.2 Solution by Dynamic Programming

Dynamic programming is a technique especially suitable for optimizing an objective function that is the summation of functions of only a subset of the set of the (mostly discrete) decision variables. So it is particularly suitable for dynamic problems in which:

$$F(\mathbf{X}_1, \ldots, \mathbf{X}_n) = \sum_{t=1}^{n} F_t(\mathbf{X}_t)$$

We will formulate problem (6.1.2) now in a more detailed way so that we can show the use of dynamic programming:

$$\max_{\mathbf{C}_1, \ldots, \mathbf{C}_n} S = \sum_{t=1}^{n} \{U_t(\mathbf{X}_t) - T_t(\mathbf{X}_t, \mathbf{C}_t) - I(\mathbf{C}_t, \mathbf{C}_{t-1})\} + S_s(\mathbf{C}_n) \qquad (6.1.5)$$

subject to:

$$A\mathbf{X}_t = \mathbf{0}, \mathbf{X}_t \geqslant \mathbf{0} \quad \text{for all } t = 1, \ldots n$$

$$G_t(\mathbf{X}_t, \mathbf{C}_t) = \mathbf{0} \qquad \text{for all } t = 1, \ldots n$$

with $I(\mathbf{C}_t, \mathbf{C}_{t-1})$—the investment costs of changing network $\mathbf{C}_{t-1}$ into network $\mathbf{C}_t$

We omitted further constraints $\mathbf{H}$ for reasons of simplicity, though it is easily possible to manage more constraints. We see now that the formulation of the problem (6.1.5) is suitable for dynamic programming. The method of dynamic programming is based on the 'Principle of Optimality' of Bellman (Bellman, 1957) which states: 'an optimal policy (series of decisions) has the property that whatever the preceding decisions or the initial state has been, the next decisions form an optimal policy for the situation that existed after the preceding decisions'. So it is not necessary to take all decisions at the same instant, but we can take them step by step in a dynamic process. With the help of the principle of optimality we can write down some recurrence relations. To derive these for the problem (6.1.5) we must define the variable S^t, the maximum surplus for the period from 1 to t, as a function of the network $\mathbf{C}_t$:

$$S^t(\mathbf{C}_t) = \max_{\mathbf{C}_1, \ldots \mathbf{C}_t} \sum_{t'=1}^{t} \{U_{t'}(\mathbf{X}_{t'}) - T_{t'}(\mathbf{X}_{t'}, \mathbf{C}_{t'}) - I(\mathbf{C}_{t'}, \mathbf{C}_{t'-1})$$

subject to:

$$A\mathbf{X}_{t'} = \mathbf{0}, \mathbf{X}_{t'} \geqslant \mathbf{0} \quad \text{for all } t = 1, \ldots, n$$
$$G_{t'}(\mathbf{X}_{t'}, \mathbf{C}_{t'}) = \mathbf{0} \qquad \text{for all } t = 1, \ldots, n \qquad (6.1.6)$$

Using this variable S^t we can write down the recurrence relations for the problem (6.1.5):

$$S^0 = 0$$

$$\mathbf{C}_0 = \text{the original network}$$

$$\begin{aligned}
S^t(\mathbf{C}_t) = \max_{\mathbf{C}_{t-1}} \{ & S^{t-1}(\mathbf{C}_{t-1}) + U_t(\mathbf{X}_t) \\
& - T_t(\mathbf{X}_t, \mathbf{C}_t) - I(\mathbf{C}_{t-1}, \mathbf{C}_t)\} \\
\text{subject to:} \quad & A\mathbf{X}_t = \mathbf{0}, \mathbf{X}_t \geqslant \mathbf{0} \\
& \mathbf{G}_t(\mathbf{X}_t, \mathbf{C}_t) = \mathbf{0}
\end{aligned} \Bigg\} \quad \text{for all } t = 1, \ldots n \quad (6.1.7)$$

$$S = \max_{\mathbf{C}_n} \{S^n(\mathbf{C}_n) + S_s(\mathbf{C}_n)\} \tag{6.1.7}$$

Using the operations defined by the recurrence relations of relationships (6.1.7) we get the optimal solution of the dynamic network optimization problem. We see that we only ever need to inspect one period at a time. Nor do we need to inspect all possible combinations of the decision variables. For all combinations which have been shown to be not profitable in an earlier period do not need to be inspected in the later periods. So the dynamic programming approach is much more efficient than an exhaustive enumeration of all

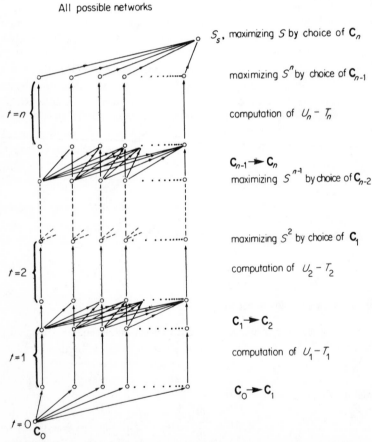

Figure 6.1.1 Illustration of the method of dynamic programming. [Reproduced by permission of Göran Bergendahl, from Figure 8.1.1 (page 98) of *Models for Investments in a Road Network* (Bonniers, Stockholm, Sweden)]

possible combinations. Furthermore all intermediate results $S^t(C_t)$ and C_t are themselves interesting because they give an optimal policy if C_t were to be the optimal solution.

The recurrence relations of relationships (6.1.7) form an example of a forward formulation of dynamic programming. This is the best method here, because the initial network is known. It is also possible to use a backward formulation. The basic recurrence relation then becomes:

$$S^t(C_t) = \{ \max_{C_{t+1}} \{ S^{t+1}(C_{t+1}) + S_t(C_t, C_{t+1}) \}$$

where $S^t(C_t)$ is the maximum surplus for the period from t to n, and

$$S_t(C_t, C_{t+1}) = U_t(X_t) - T_t(X_t, C_t) - I(C_t, C_{t+1})$$

We can use this formulation, for instance, when the final state is known and we must trace back to the initial state.

The dynamic network optimization problem is solved by dynamic programming for instance by Bergendahl (1969a, 1969b), Morlok, Thomas, et al. (1969) and Morlok, Nihan and Sullivan (1969). They all use dynamic programming to decide what changes to make to the network ($C_{t-1} \rightarrow C_t$) (the master problem) and another optimization technique—(parametric) linear programming in this case—to obtain the best values for $U_t - T_t$ (see Figure 6.1.1). Bergendahl uses a descriptive assignment (1969a) or a normative one (1969b). For 13 origins and destinations, a network consisting of 16 links and 4 time periods, he gets the optimum solution in about 5 minutes on an IBM 360/65 computer. Morlok et al. include only investments, maintenance and exploitation costs in their objective function, subject to some social constraints as maximum travel times (effectiveness of the system).

6.1.3   Solution by Other Techniques

Though dynamic programming is very suitable for the dynamic problem, it is not, of course, the only solution method possible. Roberts (1966) for instance used mixed integer programming for the dynamic problem with a normative assignment and Taborga (1966) used a method similar to that of Ridley (Section 4.3.2). We will discuss here briefly the possibility of using the methods discussed already for the static problem, namely branch and bound, heuristic programming and the stepwise assignment according to the least marginal objective function. All these methods can be used for the dynamic problem.

For the successful use of branch and bound it is important that the set of solutions can be divided into subsets of solutions and that for every subset a bound for the objective function for the solutions contained in the subset and a bound for the optimal solution can be computed in an easy way (see Section 4.3.1). Because the functions in the dynamic problem have the same properties as the functions in the static problem, namely increasing investment costs, decreasing users' costs (in general at any rate) and increasing users' benefits when projects are being implemented, and because the decision variables (dimensions per link per period) are again discrete variables, there does not seem to be any theoretical difficulty in using branch and bound for the dynamic problem. Only the number of variables and the number of constraints are much higher than for the static problem. This will, however, give rise to an important practical difficulty.

Of the heuristic techniques we will discuss the use of the method of the continuous optimal adjustment (Section 4.4.2.1) for the minimization of the costs for the whole period from 1 to n for given trip-matrices for every period t and with a descriptive assignment:

$$\min_{c_{1ij},\ldots,c_{nij}} \sum_{ij \in L} F_{ij}(x_{1ij},\ldots,x_{nij},c_{1ij},\ldots,c_{nij})$$

subject to:

$$A\mathbf{X}_t = \mathbf{0}, \mathbf{X}_t \geqslant \mathbf{0} \quad \text{for all } t = 1,\ldots n$$

$$G_t(\mathbf{X}_t, \mathbf{C}_t) = 0 \qquad \text{for all } t = 1,\ldots n \tag{6.1.8}$$

$$H_{ij}(x_{1ij},\ldots x_{nij}, c_{1ij},\ldots c_{nij}) >, =, <, 0 \quad \text{for all } ij \in L$$

(We assume the scrap value F_{sij} to be included in the F_{ij}'s.)

We solve this problem by first assigning the n trip-matrices to n starting networks $\mathbf{C}_t^{(1)}$ (first and second set of constraints). Then we assume the traffic flows $\mathbf{X}_t^{(1)}$ to be constant and solve n_L costs minimization problems with n_L third sets of constraints. We get n new networks then and assign the trip-matrices to the networks again and continue the process. So in the $(2m - 1)$th step we get the assignment problems:

find $\mathbf{X}_1,\ldots,\mathbf{X}_n$ such that

$$A\mathbf{X}_t = \mathbf{0}, \mathbf{X}_t \geqslant \mathbf{0} \quad t = 1,\ldots n$$

$$G_t(\mathbf{X}_t, \mathbf{C}_t) = 0 \qquad t = 1,\ldots n \tag{6.1.9}$$

and in the $(2m)$th step the n_L minimization problems:

$$\left.\begin{array}{l} \min_{c_{1ij},\ldots,c_{nij}} F_{ij}(x_{1ij}^{(m)},\ldots,x_{nij}^{(m)}, c_{1ij},\ldots,c_{nij}) \\[2mm] \text{subject to:} \quad H_{ij}(x_{1ij}^{(m)},\ldots,x_{nij}^{(m)}, c_{1ij},\ldots,c_{nij}) >, =, <0 \end{array}\right\} \quad \text{for all } ij \in L$$

$$\tag{6.1.10}$$

Of course, there are the same objections to the method for the dynamic problem as exist for the static problem (see Section 4.4.2.1). Finding the solution for the minimization problems (6.1.10) is not very difficult; see also Section 10.6.

Many other kinds of heuristic techniques can be used, of course, with all the advantages and disadvantages of the heuristics. Another way of using heuristics for the dynamic problem is to take into account only half of the dynamic aspects, i.e. only the past or only the future:

$$\left. \begin{array}{l} \min_{C_t} F_t(\mathbf{X}_t, \mathbf{C}_t) \\ \text{subject to } C_{t-1} \end{array} \right\} \quad \text{for } t = 1, \ldots, n \qquad (6.1.11)$$

or:

$$\left. \begin{array}{l} \min_{C_t} F_t(\mathbf{X}_t, \mathbf{C}_t) \\ \text{subject to } \mathbf{C}_{t+1} \end{array} \right\} \quad \text{for } t = n, \ldots, 1 \qquad (6.1.12)$$

Finally, we will discuss the use of the stepwise assignment according to the least marginal objective function for the dynamic costs minimization problem. The formulation of the problem is given in relationship (6.1.8). We can decompose the problem again into a master problem and n_L subproblems:

the subproblems:

$$\left. \begin{array}{l} \min_{c_{1ij},\ldots,c_{nij}} F_{ij}(x_{1ij}, \ldots, x_{nij}, c_{1ij}, \ldots, c_{nij}) \\ \text{subject to: } H_{ij}(x_{1ij}, \ldots, x_{nij}, c_{1ij}, \ldots, c_{nij}) >, =, <0 \end{array} \right\} \text{ for all } ij \in L \quad (6.1.13)$$

resulting in the functions $F_{ij}^{\min}(x_{1ij}, \ldots, x_{nij})$.

the master problem:

$$\min_{x_{1ij}, \ldots, x_{nij}} \sum_{ij \in L} F_{ij}^{\min}(x_{1ij}, . . , x_{nij})$$

subject to: $\quad A\mathbf{X}_t = \mathbf{0}, \mathbf{X}_t \geqslant \mathbf{0} \quad$ for all $t = 1, \ldots, n$ $\qquad (6.1.14)$

The subproblems (6.1.13) do not give any mathematical difficulties (the use of dynamic programming again is very advantageous here (see also Section 10.6) and for the master problem we can state again the conditions for the optimal solution in the case of convex functions:

$$\left. \begin{array}{l} \text{if } x_{tij}^{ab} > 0 \quad \text{then } F_t^{\prime ai} + F_{tij}^{\prime} = F_t^{\prime aj} \\ \text{if } F_t^{\prime ai} + F_{tij}^{\prime} > F_t^{\prime aj} \quad \text{then } x_{tij}^{ab} = 0 \end{array} \right\} \begin{array}{l} \text{for all } ab \in P \\ \text{for all } ij \in L \\ \text{for all } t = 1, \ldots n \end{array} \quad (6.1.15)$$

in which $\quad F_{tij}^{\prime} = \left(\dfrac{\partial F_{ij}^{\min}}{\partial x_{tij}} \right)_{\substack{x_{1ij} = x_{1ij}^* \\ \vdots \\ x_{nij} = x_{nij}^*}}$.

We can use now a similar stepwise assignment technique to that used in Section 5.4 in which we first assign a part α_{11} of the trip matrix of period 1 to the network then a part α_{21} of the trip-matrix of period 2 and so on. The criterion for the mth step for the assignment of the trip-matrix of period t then becomes:

$$\{F_{ij}^{\min}(x_{1ij}^{\alpha_{1}m}, \ldots, x_{t-1\,ij}^{\alpha_{t-1}m}, x_{tij}^{\alpha_{t}m-1} + \Delta x^{tm}, x_{t+1\,ij}^{\alpha_{t+1}m-1}, \ldots, x_{nij}^{\alpha_{n}m-1})$$
$$- F_{ij}^{\min}(x_{1ij}^{\alpha_{1}m}, \ldots, x_{t-1\,ij}^{\alpha_{t-1}m}, x_{tij}^{\alpha_{t}m-1}, x_{t+1\,ij}^{\alpha_{t+1}m-1}, \ldots, x_{nij}^{\alpha_{n}m-1})\} \tag{6.1.16}$$

Of course it is also possible to start with period n or any other period. The same difficulties we had in Section 5.4 arise here and we must try to get the best result by choosing the parameters of the optimization process in a sensible way. These parameters are, for the dynamic problem:

(a) the number of steps per period;
(b) the order of succession of the assignment processes;
(c) α_{tm};
(d) Δx^{tm}.

6.1.4 Optimization for One Time Period but with Fluctuating Traffic

In Section 6.1.1. we spoke about the optimization problem for many networks and many situations at different times. But we also have the problem of one network for many situations: the traffic is different every minute of the day, every day of the week and so on; but the traffic flows still use the same transport network. This too is a dynamic problem very similar to the problem of optimization over time. For the surplus maximization, descriptive case, the problem can be stated as follows (comparable with relationships (6.1.1) and (6.1.2)):

$$\max_{\mathbf{C}} \int_{t_0}^{t_e} S_t(\mathbf{X}(t), \mathbf{C})\, dt$$

subject to:

$$\left. \begin{array}{l} A\mathbf{X}(t) = \mathbf{0}, \mathbf{X}(t) \geqslant \mathbf{0} \\[4pt] \mathbf{G}_t(\mathbf{X}(t), \mathbf{C}) = \mathbf{0} \\[4pt] \mathbf{H}(\mathbf{X}(t), \mathbf{C}) >, =, < \mathbf{0} \end{array} \right\} \quad \text{for } t_0 \leqslant t \leqslant t_e \tag{6.1.17}$$

or for the discrete case:

$$\max_{\mathbf{C}} \sum_{t=1}^{n} S_t(\mathbf{X}_t, \mathbf{C})$$

subject to:

$$\left. \begin{array}{l} A\mathbf{X}_t = \mathbf{0}, \mathbf{X}_t \geqslant \mathbf{0} \\[4pt] \mathbf{G}_t(\mathbf{X}_t, \mathbf{C}) = \mathbf{0} \\[4pt] \mathbf{H}(\mathbf{X}_t, \mathbf{C}) >, =, < \mathbf{0} \end{array} \right\} \quad \text{for all } t = 1, \ldots, n \tag{6.1.18}$$

The use of dynamic programming to solve this problem (6.1.18) is not very profitable because of the presence of only one (set of) decision variables. All methods discussed in Section 6.1.3 can be used. For the continuous optimal adjustment the assignment problem for problem (6.1.18) is identical with problem (6.1.9) from Section (6.1.3) and the minimization problem is slightly different from problem (6.1.10). The small difference arises because in the case of one network for fluctuating traffic there is only one decision variable c_{ij} instead of c_{1ij}, c_{2ij} and so on for the real dynamic problem. For the stepwise assignment according to the least marginal objective function, the master problem for problem (6.1.18) is again identical with the master problem (6.1.14), whereas the subproblems are slightly different from the subproblems (6.1.14), because, again, only one decision variable c_{ij} is involved.

6.2 OPTIMIZATION OF NETWORKS FOR MORE TRAVEL MODES

6.2.1 Statement of the Problem

In the preceding two chapters we assumed only one travel mode. In this section we will investigate what difficulties arise in optimizing networks for two or more travel modes. We can state the objective function for the surplus maximization as follows:

$$\max \sum_{ab \in P} U^{ab} - \sum_{\substack{m \\ ij \in L}} (T_{mij} + i_{mij}) \qquad (6.2.1)$$

with:

T_{mij}—total users' costs on link ij by travel mode m.
i_{mij}—total investments in link ij of travel mode m.

The objective function does not look very different from the objective function used for the unimodal network optimization; but it is. To make this clear we must remember some of the things we discussed in Section 3.1 and on the other hand anticipate a little the things to be discussed in Chapter 9. As we said in Section 3.1 there exists a strong relationship between the land use pattern and the transport network. Therefore we must treat the physical planning and the transportation planning as one large optimization problem. However, this turned out to be very, if not too, difficult. Still the interdependency between land use and infrastructure exists. And for the multimodal network optimization the totally different relationships between transport and land use for the different travel modes are very important. By way of illustration we can just mention:

(a) the necessity of heavy transport flows to make the use of public transport worthwhile, but on the other hand the ability to transport many persons in a short time and relatively cheaply when there are heavy transport flows;

(b)   the different use of space by car traffic (including parking) and rail transport; in urban situations, the totally different layout of car-oriented and public-transport-oriented towns is striking.

For an ample discussion of the interdependency between land use and the use of the different travel modes see, for instance, Buchanan (1963) and van Dam (1971).

Of course the coherence between land use and infrastructure also exists for networks for just one travel mode. But here we know that quite often the different links and traffic flows on them possess comparable effects on the land use. So it might be possible to neglect them in a first attempt (see also Chapter 8). For the multimodal optimization this neglecting would not appear possible, because the above-mentioned factors are quite often the essential factors in the choice between investments in highways or railways. So these things must be included in the objective function now, a very difficult process.

Another difficulty for the objective function is presented by the users' costs T_{mij}. Let us focus here—again as an illustration—on the travel time costs. We will leave the discussion on the travel time evaluation to Section 9.2.1, but we do have to say something about it here.* The important thing is that we cannot use the same value for the travel time for the different transport modes (Steenbrink, 1970b).

In Section 2.1.2 we said that we might describe the behaviour of the traveller as the result of an individual optimization process:

$$\max_{m,p,b} \{u^{ab} - t^{mpab}\}$$

Assuming that the traveller does travel from a to b and ignoring the route choice, the individual optimization problem becomes:

$$\min_{m} t^{m} \quad (a, b \text{ and } p \text{ are omitted}) \tag{6.2.2}$$

Assuming only two travel modes (auto and train) we get:

$$\left.\begin{array}{ll} \text{the traveller goes by auto} & \text{if } t^{\text{auto}} < t^{\text{train}} \\ \text{the traveller goes by train} & \text{if } t^{\text{auto}} > t^{\text{train}} \\ \text{the traveller may go by either mode} & \text{if } t^{\text{auto}} = t^{\text{train}} \end{array}\right\} \tag{6.2.3}$$

Now we assume the users' costs to consist of an objective travel time t' multiplied by a personal, travel time evaluation **k** (we use bold letters here

---

* For the ideas worked out in this section I am much indebted to Dr. R. Hamerslag.

to indicate stochastic variables as opposed to deterministic variables):

$$\mathbf{t}^{auto} = \mathbf{k}_{auto} t'^{auto}$$
$$\mathbf{t}^{train} = \mathbf{k}_{train} t'^{train}$$

(6.2.4)

if $\mathbf{k}_{auto}$ and $\mathbf{k}_{train}$ were the same ($\mathbf{k} = \mathbf{k}_{auto} = \mathbf{k}_{train}$) everybody would go by auto or by train, for:

$$kt'^{auto} \leqslant kt'^{train} \quad \text{or} \quad kt'^{auto} \geqslant kt'^{train}$$

Because we very seldom see such an all-or-nothing modal split we know that $\mathbf{k}_{auto}, \mathbf{k}_{train}$ and also $\mathbf{k} = \mathbf{k}_{auto}/\mathbf{k}_{train}$ are different for everybody.

We try now to find the distribution of the variable κ. For its derivation we continue our treatment of the modal split (6.2.3). For the people who go by car it holds that:

$$\mathbf{k}_{auto} t'^{auto} \leqslant \mathbf{k}_{train} t'^{train}$$

or:

$$\mathbf{k} = \frac{\mathbf{k}_{auto}}{\mathbf{k}_{train}} \leqslant \frac{t'^{train}}{t'^{auto}} = \tau$$

(6.2.5)

The share of car travellers A is then:

$$A = \int_0^\tau F(\kappa)\, d\kappa$$

(6.2.6)

so:

$$F(\kappa) = \frac{dA}{d\kappa} \quad \text{or} \quad \frac{dA}{d\tau}$$

(6.2.7)

From these relationships (6.2.6) and (6.2.7) we see that the distribution of κ is known if A is known as a function of τ. But the relationship between modal split and travel time ratio is one of the most popular modal split formulae, so the distribution of κ is known in many cases. So to get the evaluated travel time costs we cannot write:

$$\kappa x (At'^{auto} + Tt'^{train})$$

(6.2.8)

but we have to write, for instance:

$$x \left\{ Tk_{train} t'^{train} + k_{train} t'^{auto} \int_0^{\tau^*} \frac{dA}{d\kappa} \kappa\, d\kappa \right\}$$

(6.2.9)

in which k_{train} is some average travel time evaluation for the travellers by train and τ^* is the actual travel time ratio.

Having discussed some aspects of the objective function we turn now to the decision variables and constraints. For the unimodal optimization the dimensions of the links and sometimes the traffic flows are the decision variables. For a correct multimodal optimization we must use more decision variables. Besides the dimensions of the links (number of lanes or number of tracks) there are the parking facilities for private transport and the number and positions of stations and stops, the speed and frequency of the trains and buses and so on, for public transport.

For the constraints we can have the whole transportation model again or a given trip-matrix and also the descriptive or the normative case. Using a given trip-matrix it is clear that we mean a given trip-matrix for the total transport. The split into the different travel modes is an essential part of the multimodal network optimization. Working with a normative model, i.e. a normative modal split and normative assignments, does not look too realistic. At the assignment we can assume that the route choice is influenced fairly easily by traffic management or pricing systems (see Section 5.5), but this seems to be much more difficult for the modal choice. Furthermore, a change in the choice of travel mode is much more far-reaching for the traveller than a change in the route choice.

Having seen the difficulties of defining the objective function and the increasing number of variables and constraints we can understand that very few attempts have been made to solve the multimodal network optimization problem. The only work I know is that of Morlok et al. (Hay, Morlok and Charnes, 1965, Morlok, Nihan and Sullivan, 1969 and Morlok, Thomas et al., 1969). Their objective function is the minimization of investments and exploitation costs, subject to some effectiveness constraints, such as a maximum travel time for every relation. In this way they shift the difficulty from defining the objective function to defining the constraints. They then introduce rather arbitrary (say political) values for the requested effectiveness level, which have, of course, a strong influence on the solutions.

One last thing we have to mention in this section is that in many cost/benefit studies for infrastructure projects the effect a certain project relevant to one mode has on the other modes is included in the computation as an external effect (see for instance Werkgroep 'Spoorlijn Amsterdam–Den Haag', 1969).

6.2.2   Possible Solution Methods

In this section we will discuss the possibilities of the solution methods discussed in the preceding chapters for the multimodal network optimization, assuming that it has been possible to state the objective function in a proper way. Branch and bound in (Section 4.3) seems theoretically very suitable—at least for (very) small networks—because it can manage very complicated objective functions and constraints. The only requirement is that it be easily possible to define upper and lower bounds for the optimal solution or a particular set of solutions. This seems easy to be done, because the investment costs will be increasing as projects are being implemented and the users' costs (for instance travel time costs evaluated in the right way) will be decreasing as projects are being implemented. The users' benefits will also increase in that case. So branch and bound should be very suitable; if it is not there should be a restriction on the size of the problem.

With the use of heuristic methods (Section 4.4) we must be even more careful than we had to be in the unimodal case. We will give an example here of the use of the continuous optimal adjustment (Section 4.4.2.1) for the multimodal optimization (Steenbrink, 1970a). We take only one relation and a network consisting of a rail connection and a motor road. The problem is formulated as follows:

$$\min_{c_a, f_t, v_t} F = xk(At_a + Tt_t) + i_a + i_t + e_t$$

subject to:

$$x = 10{,}000$$

$$A = \begin{cases} \left(1 - \dfrac{t_a}{2 \cdot 5t_t}\right) & \text{if } t_a \leqslant 2 \cdot 5t_t \\ 0 & \text{if } t_a > 2 \cdot 5t_t \end{cases}$$

$$T = 1 - A$$

$$t_a = 95\left\{0 \cdot 5 + 0 \cdot 25\left(\frac{Ax}{c_a}\right)^2\right\}$$

$$i_a = 95\sqrt{2{,}000c_a}$$

$$t_t = 95/v_t + 20/f_t + 2$$

$$i_t = 95{,}000v_t$$

$$e_t = 9{,}500f_t v_t^2$$

$$k = 3 \tag{6.2.10}$$

with:

c_a—dimension of the motor road
f_t—frequency of the train (number of trains per hour)
v_t—speed of the train
e_t—exploitation costs for the train

We start the process with some modal split ($A^{(1)}$, $T^{(1)}$) and then minimize the objective function, ignoring the modal split constraints:

$$\frac{\partial F}{\partial c_a} = \frac{d(xkA^{(1)t}{}_a + i_a)}{dc_a} = 0 \tag{6.2.11}$$

$$\frac{\partial F}{\partial f_t} = \frac{\partial (xkT^{(1)}t_t + i_t + e_t)}{\partial f_t} = 0$$

$$\frac{\partial F}{\partial v_t} = \frac{\partial (xkT^{(1)}t_t + i_t + e_t)}{\partial v_t} = 0 \qquad (6.2.12)$$

Relationship (6.2.11) yields an optimal value for the dimension of the motor road: $c_a^{(1)}$ and the relationships (6.2.12) yield the optimal values for the speed and the frequency of the train: $v_t^{(1)}$ and $f_t^{(1)}$; everything for the first estimate for the modal split.

We use the results $c_a^{(1)}$, $f_t^{(1)}$, $v_t^{(1)}$ to compute a new modal split, ignoring the optimization now. This process is continued till no further changes in the variables appear. We have tried this process for several starting modal splits:

0·9999/0·0001, 0·95/0·05, 0·90/0·10, 0·80/20 . . . up to 0·0001/0·9999.

Except for the starting modal split $T = 0.9999$, $A = 0.0001$, all starting solutions converged to $T = 0.4404$, $A = 0.5596$ with $F = 2,362,369$. Only the starting solution $T = 0.9999$ and $A = 0.0001$ converged to the real optimum solution $T = 1$, $A = 0$ with $F = 1,734,623$, that is about 25 per cent less. So it is obvious that this solution method does not give the optimal solution (see further Section 4.4.2.1). Still this process does seem to occur in reality (think of the decline of the railways in the United States). Of course it will be possible and perhaps often necessary to use heuristic techniques. However, we must be very careful in using them.

The method of the stepwise assignment according to the least marginal objective function (Chapter 5) seems to be ruled out, for the problem as a whole at any rate, by the necessity of using a normative modal split.

As a last remark we will mention the possibilities of decomposition (see Section 4.5). A decomposition into subproblems for the different modes and a master problem to define the overall policy seems very plausible.

6.3 SOME REMARKS ON THE OPTIMAL NETWORK STRUCTURE

6.3.1 Some Simple Network Structure Optimizations

When dealing with the optimal network structure, we have the same objective function as for the 'normal' network optimization problem, defined in Section 4.1, but more decision variables. In addition to the traffic flows and the dimensions of the links, the number and the location of the links and of the nodes, except of course the origins and destinations, are also decision variables. The problem is most easily treated as costs minimization in the normative case:

$$\min_{N, L, x_{ij}, c_{ij}} \sum_{ij \in L} \{i_{ij}(c_{ij}) + x_{ij}t_{ij}(x_{ij}, c_{ij})\} \qquad (6.3.1)$$

subject to:

the number and location of the origins and destinations;
a fixed trip-matrix;
the network constraints.

For small problems one tries to construct solutions in a geometrical way using compasses and ruler or one states an analytical optimization problem with the number of links and nodes given and the coordinates of the nodes as decision variables (Beckmann 1960, Beckmann 1967, Werner 1966, Gilbert 1967). The problem is of interest not just in theory, but also in practice in the planning of towns and regions (Smeed 1963, Tanner 1967, Smeed 1971).

We will mention here some simple problems. When $i_{ij} = 0$ and t_{ij} is only a function of the geographical length of link ij, the solution of problem (6.3.1) is very simple. Direct connections for all relations form the optimal solution. This is generally true when the total objective function is a linear function of the traffic flows multiplied by the geographical lengths of the links. On the other hand when $t_{ij} = 0$ and i_{ij} is only a function of the geographical length of the link we get the problem of finding the 'Steiner minimal tree', a graph which can be constructed by ruler and compasses (Melzak, 1961) (N.B. when the number and location of the nodes are given this is the well-known problem in Operations Research of the minimum spanning tree).

When i_{ij} and $x_{ij}t_{ij}$ are both contained in the objective function we find that 'bundling' has some advantages, the question being how far this bundling must go. Figure 6.3.1 shows the simplest problem with two traffic flows x^{12}

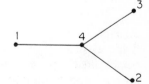

Figure 6.3.1   Optimal location of node 4 (traffic flows x^{12} and x^{13})

and x^{13}; the question is the location of the node 4 where the road splits. When we take for i_{ij} and t_{ij} linear functions of the geographical length it is possible to state and solve an analytical minimization problem for the location of node 4. This is accomplished as follows:

Give node 1 the coordinates $(0, 0)$ and the nodes 2, 3 and 4 the coordinates respectively (ξ_2, η_2), (ξ_3, η_3) and (ξ, η). Suppose $i_{ij} = il_{ij}$ (with l_{ij} length of ij)

and $t_{ij} = kl_{ij}$. The problem is:

$$\min_{\xi,\eta} F = \{i + k(x^{12} + x^{13})\}\sqrt{\xi^2 + \eta^2}$$
$$+ (i + kx^{12})\sqrt{(\xi - \xi_2)^2 + (\eta - \eta_2)^2}$$
$$+ (i + kx^{13})\sqrt{(\xi - \xi_2)^2 + (\eta - \eta_3)^2}$$

$$\frac{\partial F}{\partial \xi} = 0 \quad \text{and} \quad \frac{\partial F}{\partial \eta} = 0 \quad \text{gives the solution.}$$

Using trigonometry it is not too difficult to solve these equations and to get relationships between the cosines of the angles in node 4 and the parameters $(i, k,$ traffic flows) of the problem. Further Werner (1966) shows that this point does indeed give the minimum for the objective function. Moreover it is possible to construct the position of node 4 by ruler and compasses (see Werner 1966, page 41).

The problem becomes more complicated when there are also traffic flows between 2 and 3, from 2 to 1 and from 3 to 1. The general shape of the solution is shown in Figure 6.3.2(a), but this shape can degenerate to the

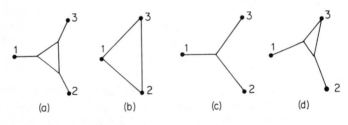

Figure 6.3.2   Optimal structure of the network between the nodes 1, 2 and 3 for the traffic flows $x^{12}, x^{21}, x^{13}, x^{31}, x^{23}$ and x^{32}.

Δ-shape of Figure 6.3.2(b), the Y-shape of Figure 6.3.2(c) or an intermediate shape as in Figure 6.3.2(d). The direct analytical solution is much more complicated here than for the problem of Figure 6.3.1. We can solve this problem too by iteratively solving a system of three two-flow problems (assuming x^{ij} and x^{ji} to be one flow). It is always possible to reduce complex situations with many relations to the form of Figure 6.3.2 and to solve complex situations by solving repeatedly the more simple situation. We must however be careful to take into account every possible topological structure.

Another simple and well-known problem is that of a destination beside an existing road and the question is where to leave the existing road and to build a new one to that destination (Figure 6.3.3). The analytical solution

Figure 6.3.3   Optimal location of node 3
(traffic flow x^{12})

to this problem proceeds as follows:

Put the ξ-axis along the existing road and node 1 into the origin and give the nodes 2 and 3 the coordinates respectively (ξ_2, η_2) and $(\xi, 0)$. Suppose $t_{ij} = kl_{ij}$, $i_{ij} = 0$ for the existing road and $i_{ij} = il_{ij}$ for the new road. With a traffic flow x^{12} the problem becomes:

$$\min F = kx^{12}\xi + (i + kx^{12})\sqrt{\eta_2^2 + (\xi_2 - \xi)^2}.$$

$$\frac{dF}{d\xi} = 0 \quad \text{gives the solution} \quad \cos \phi = \frac{kx^{12}}{i + kx^{12}}.$$

This problem can also be easily solved geometrically. Of course it would be possible here to state and solve similar but more complicated problems.

6.3.2   Optimal Alignment

The problems discussed in the preceding sections are in the first place of theoretical interest, although one can try to learn something of the optimal structure in general and related topics. For real situations (say the road network for a region or a country) the cost functions, are, and therefore the problem is, much more complicated. On a very small but also very detailed scale we meet the same kind of problem in defining the optimal horizontal and/or vertical alignment.

The problem of the horizontal and vertical alignment is the problem of the optimal location of a road between two points when all costs (investments and users' costs) are known as functions of the location. There is a number of possible ways to state and solve the problem. It is treated most easily in a dynamic way, that is the best 'route' is sought to go from one point to the other one. Howard *et al.* (1968), for instance, use the following formulations which are suitable for the application of the calculus of variations:

$$\min_{\xi(t),\eta(t),\zeta(t)} \int_{t_1}^{t_2} f\{\xi(t), \eta(t), \zeta(t), \dot{\xi}(t), \dot{\eta}(t), \dot{\zeta}(t)\} \, dt$$

subject to:

$$\xi(t_1) = \xi_1$$
$$\eta(t_1) = \eta_1$$
$$\zeta(t_1) = \zeta_1$$
$$\xi(t_2) = \xi_2 \qquad\qquad (6.3.2)$$
$$\eta(t_2) = \eta_2$$
$$\zeta(t_2) = \zeta_2$$

with

     ξ, η, ζ—coordinates

         t—time

     $\dot{\xi}, \dot{\eta}, \dot{\zeta}$—derivatives to t

For f we choose $F(\xi, \eta, \zeta)\sqrt{\dot{\xi}^2 + \dot{\eta}^2 + \dot{\zeta}^2}$; in which $F(\xi, \eta, \zeta)$ are the total costs at point ξ, η, ζ. So we assume that the total costs are known for every point between 1 and 2. Using the calculus of variations (Pontryagin, 1961) we get the Euler–Lagrange conditions for the coordinates of the optimal trajectory:

$$\frac{d}{dt} \frac{\dot{\xi}F(\xi, \eta, \zeta)}{\sqrt{\dot{\xi}^2 + \dot{\eta}^2 + \dot{\zeta}^2}} = \sqrt{\dot{\xi}^2 + \dot{\eta}^2 + \dot{\zeta}^2}\,\frac{\partial F}{\partial \xi}$$

$$\frac{d}{dt} \frac{\dot{\eta}F(\xi, \eta, \zeta)}{\sqrt{\dot{\xi}^2 + \dot{\eta}^2 + \dot{\zeta}^2}} = \sqrt{\dot{\xi}^2 + \dot{\eta}^2 + \dot{\zeta}^2}\,\frac{\partial F}{\partial \eta} \qquad (6.3.3)$$

$$\frac{d}{dt} \frac{\dot{\zeta}F(\xi, \eta, \zeta)}{\sqrt{\dot{\xi}^2 + \dot{\eta}^2 + \dot{\zeta}^2}} = \sqrt{\dot{\xi}^2 + \dot{\eta}^2 + \dot{\zeta}^2}\,\frac{\partial F}{\partial \zeta}$$

Differentiation and some working out yields the Optimum Curvature Principle of Howard, Bramnick and Shaw (1968), which gives relations for the optimum curvature of a route.

It is also possible to use a more discrete formulation. We can, for instance, put the η-axis through the two nodes, 1 and 2, to be connected. Then we assume that we go in n steps from node 1 with η_1, to node 2 with η_2. More-ever we assume that we know the costs of going from (ξ_i, η_i, ζ_i) to $(\xi_{i+1}, \eta_{i+1}, \zeta_{i+1})$ as a function of $\xi_i, \zeta_i, \xi_{i+1}$ and ζ_{i+1}, at least for some locations and possible transitions from i to $i + 1$. Then we get the following formulation:

$$\min_{\xi_2, \zeta_2, \ldots, \xi_{n-1}, \zeta_{n-1}} \sum_{i=1}^{n-1} F_i\{\xi_i, \zeta_i, \xi_{i+1}, \zeta_{i+1}\} \tag{6.3.4}$$

The problem formulated in this way can be solved very easily by dynamic programming, as is done for instance in the civil-engineering software-system GCARS (Turner, 1971).

Other optimization techniques can also be used of course, e.g. other mathematical programming techniques or direct application of a shortest-path algorithm, see for instance the proceedings of two symposiums of the Planning & Transport Research & Computation (1969 and 1971).

6.3.3 Indicators for the Structure of a Graph

As results of the optimization of network structure we get the network and the value of the different elements of the objective function. For the further evaluation and mutual comparison of results it is sometimes advantageous to have some general indicators to measure certain aspects of the spatial structure of the networks. Such indicators have been developed in the theory of graphs. One can find an extensive treatment of them for instance in Harrison and Marble (1962), Kansky (1963), Morlok (1967) and Haggett and Chorley (1969). Below we will mention briefly some of these indicators.

To measure the circuity in a network Morlok (1967) proposes the following indicator:

$$\frac{\sum_{ab \in P} l^{ab}}{\sum_{ab \in P} l^{*\,ab}}$$

with:

l^{ab} —the length of the shortest path in the network from a to b;
$l^{*\,ab}$—the distance from a to b as the crow flies

Harrison and Marble (1962) propose two indices to measure the connectivity of a graph. The first one is the α-index, the ratio of the actual number of circuits in a planar graph and the maximum possible number. The second one is the γ-index, the ratio of the actual number of links and the maximum possible number of links in a planar network. These indices are also defined for non-planar networks. Related to the α- and the γ-index is the β-index, suggested by Kansky (1963). The β-index is just the ratio of the number of links and nodes. So we get:

$$\alpha = \frac{n_L - n_N + 1}{2n_N - 5}$$

$$\beta = \frac{n_L}{n_N}$$

$$\gamma = \frac{n_L}{3(n_N - 2)}$$

There exist many other similar indicators for the structure of a graph.

REFERENCES

Beckmann, M. J. (1960). Principles of optimum location for a transportation network. Paper presented at *The NAS–NRC Symposium on Quantitative Problems* Geography.

Beckmann, M. J. (1967). *Principles of Optimum Location for Transportation Network* Studies in Geography 13. Northwestern University, Evanston, Ill.

Bellman, R. E. (1957). *Dynamic Programming*. Princeton University Press, Princeton N.J.

Bergendahl, G. (1969a). *Models for Investments in a Road Network*. Bonniers, Stockholm.

Bergendahl, G. (1969b). A combined linear and dynamic programming model for interdependent road investment planning. *Transportation Research*, **3**, No. (July).

Buchanan, C. D. *et al.* (1963). *Traffic in Towns*. Her Majesty's Stationery Office London.

Dam, F. van (1971). *Keuze-elementen in het Stedelijk Personenvervoer*, Nederland Vervoerswetenschappelijk Instituut, Rotterdam.

Garrison, W. L., and Marble, D. F. (1962). *The Structure of Transportation Network* The Transportation Centre, Northwestern University, Evanston, Ill.

Gilbert, E. N. (1951). Minimum cost communication networks. *The Bell System Technical Journal* (November).

Haggett, P., and Chorley, R. J. (1965). *Network Analysis in Geography*, Edward Arnold London.

Hay, G. A., Morlok, E. K., and Charnes, A. (1965). *Toward Optimal Planning of Two-mode Urban Transportation System: a Linear Programming Formulation* The Transportation Center and The Technological Institute, Northwestern University, Evanston, Ill.

Howard, B. E., Bramnick, Z., and Shaw, J. F. B. (1968). Optimum curvature principle in highway routing. *Journal of the Highway Division, Proceedings of the American Society of Civil Engineers*, **94**, No. HW 1 (June).

Kansky, K. J. (1963). *The Structure of Transportation Networks*. Research Paper no. 8 University of Chicago, Department of Geography, Chicago.

Melzak, Z. A. (1963). On the problem of Steiner. *Canadian Mathematics Bulletin*, **4**, 14

Mori, Y. (1968). *A Highway Investment Planning Model: An Application of Dynamic Programming*, M.S. Thesis, M.I.T., Department of Civil Engineering, Cambridge Mass.

Morlok, E. K. (1967). *An Analysis of Transport Technology and Network Structure*, The Transportation Center, Northwestern University, Evanston Ill.

Morlok, E. K., Nihan, N. L. and Sullivan, R. F. (1969). *A Multiple-mode Network Design Model*. The Transportation Center, Northwestern University, Evanston, Ill.

Morlok, E. K., Thomas, E. N., *et al.* (1969). *The Development of a Geographic Transportation Network Generation and Evaluation Model*. The Transportation Center, Northwestern University, Evanston, Ill.

Neufville, R. de (1969). Efficient highway staging over time by dynamic programming. *Proceedings of the Planning & Transport Research & Computation. Symposium on 'Cost Models and Optimization in Road Location, Design and Construction'*, London (June).

Planning and Transport Research & Computation (1969). *Proceedings of the PTRC-symposium: 'Cost Models and Optimisation in Road Location, Design and Construction'*, London (25–27 June).

Planning & Transport Research & Computation (1971). *Proceedings of the PTRC-symposium: 'Cost Models and Optimisation in Road Location and Design'*, London (8–11 June).

Pontryagin, L. S. (1961). *Mathematical Theory of Optimal Processes*, John Wiley, New York.

Roberts, P.O. (1966). *Transport Planning: Models for Developing Countries*, Ph.D. Thesis, Department of Civil Engineering, Northwestern University, Evanston, Ill.

Smeed, R. J. (1963). Road development in urban areas: the effect of some kinds of routing systems on the amount of traffic in the central areas of towns, *Journal of the Institution of Highway Engineers*.

Smeed, R. J. (1971). Invloeden van het ontwerp van het wegennet op de gemiddelde reisafstand en de verkeersdistributie in de stad. *Verkeerstechniek*, **22**, No. 5 (May).

Steenbrink, P. A. (1970a). *Optimalisering van de Infrastruktuur: Notitie 3. Gezamenlijke Optimalisering van Openbaar en Partikulier Vervoer*, Internal note of the Nederlandse Spoorwegen, OP2/239/41 (221)M, Utrecht.

Steenbrink, P. A. (1970b). *Optimalisering van de Infrastruktuur: Notitie 6. De Doelstellingsfunktie voor Gezamenlijke Optimalisering van Partikulier en Openbaar Vervoer. Enige Aantekeningen bij de Keuze van de Vervoerswijze*, Internal note of the Nederlandse Spoorwegen, OP2/239/41 (368)M, Utrecht.

Taborga, P. N. (1966). *A Model to Study on Optimal Transportation Policy in Chile*, Report R66-8, M.I.T., Department of Civil Engineering, Cambridge, Mass.

Tanner, J. C. (1967). *Layout of Road Systems on Plantations*, Road Research Laboratory Report LR 68, Crowthorne.

Turner, A. K. (1971), GCARS. An approach to computer aided route selection. *Proceedings of the Planning & Transport Research & Computation Symposium: 'Cost Models and Optimization in Road Location and Design'*, London (June).

Werkgroep 'Spoorlijn Amsterdam–Den Haag' (1969). *Eindrapport Schiphollijn*, Staatsuitgeverij, 's-Gravenhage.

Werner, Ch. (1966). *Zur Geometrie von Verkehrsnetzen*, Dietrich Reimer, Berlin.

7

Shortest-path Algorithms

7.1 INTRODUCTION

7.1.1 Justification of this Chapter

The length of a path can be expressed in travel distance, but it can also be expressed in users' costs, travel time, marginal objective function etc. The lengths of the shortest paths and those expressed in users' costs must be known for the computations in the transportation model (see Section 2.1.2) and for the evaluation of the objective function (see Section 3.2.2). Furthermore, in order to know where the traffic appears, in what road to invest etc., the shortest routes themselves must be known. We have seen in Section 2.1.4.2 that when we assume that the users' costs are a function of the traffic flow (as in the case of congestion) we even need to compute repeatedly all shortest paths and the users' costs on them to get an assignment of the traffic to the network (capacity restraint).

For almost all network optimization techniques we also need to compute the objective function, and consequently the total users' costs and the routing many times (recall, for instance, the branch and bound techniques). Altogether this means very many computations of the shortest paths. For the method of the stepwise assignment according to the least marginal objective function—a method especially devised to overcome the difficulty of repeated computation of the shortest paths—with four steps, about 70 per cent of the computation time is still used by the path-finding process and about 1 per cent by the assignment, thus leaving only about 15 per cent for the other computations including the reading, testing and printing of the input and the output. When the stepwise assignment according to the least marginal objective function is followed by a descriptive assignment and an optimal adjustment more than 90 per cent of the computing time is used by the path finding and assignment process.

Unfortunately, the computation of the shortest path takes a relatively very long computing time: for the road network for the Dutch Integral Transportation Study consisting of about 2,000 nodes and 6,000 links and with 351 origins and destinations one shortest path computation for all origins and destinations followed by an assignment takes about 12 minutes on an IBM 360/65 computer, using an algorithm based on the algorithm of Moore/Ford/Bellman (Section 7.2.1) written in FORTRAN IV. So it has become clear now that the use of a good shortest-path algorithm is extremely important and may even be an important factor in the total costs of a whole transportation study. On the other hand, because study budgets are often more or less fixed in advance, the quality of the shortest-path algorithm used influences the quality of the transportation study a great deal. So it is easy to understand why so much literature exists on this subject (see for instance Dreyfus 1969, Jansen 1970, Jansen 1971, Ribbeck 1971a, Ribbeck 1971b). We will give some main lines of this literature in this chapter.

7.1.2 Statement of the Problem

In Section 1.2.1 we defined the shortest path from one node to another. The shortest path from a to b is that path (as defined in Section 1.2.1) from a to b among all possible paths from a to b which has the smallest value for the 'length'. The length of a path is the sum of the lengths of all links that are contained in the path.

In this chapter we will use the following notation:

$d*^{ab}$—length of the shortest path from a to b

d^{ab} —length of the shortest path from a to b actually found at a certain moment of the computation.

Furthermore we will sometimes call d^{ab} the distance of node b; we mean then the length of the path from a to b. We also defined the length of the shortest path in a recursive way:

and

$$\left. \begin{aligned} d*^{aj} &= \min_{\substack{i \\ (ij \in L)}} (d*^{ai} + d_{ij}) \\ d*^{aa} &= 0 \end{aligned} \right\} \tag{7.1.1}$$

This definition also forms the basic operation in the algorithms to compute this shortest path. The shortest path itself is defined by the chain $ai_1i_2i_3 \ldots i_nj$ in which $i_1, i_2, \ldots, i_n$ are the solutions of the minimization problem (7.1.1).

There is a number of possible formulations for the statement of the shortest-path problem:

a) find the shortest path *from one node to one* other node;

b) find the shortest paths *from one node to all* other nodes;

c) find the shortest paths *between all* nodes.

Of course there also exist intermediate formulations, such as the problem of the shortest paths between all origins and destinations; and it is also possible to use n_N times the problem of finding the shortest path from one node to all other nodes to get the shortest paths between all nodes or to use an algorithm for finding the shortest path from one node to all other nodes, just to get one special shortest path—two things very often done in practice.

Besides the problem of finding the shortest path there is the problem of finding the nth shortest path (see Hoffman and Pavley, 1959) and of finding the shortest path consisting of no more than n links (for instance an intermediate result of some so-called treebuilding algorithms). Furthermore there are some typical extensions, of interest to the traffic engineer, such as working with turn penalties and prohibitions (see for instance Kirby and Potts, 1969) and for public transport networks, including waiting and changing times (see for instance Hamerslag, 1970 and Le Clercq, 1972). These algorithms generally use the 'normal' well-known shortest-path algorithms, but for a modified (extended) network. Also some algorithms can manage positive and negative link-lengths (using an indicator for loops) while others can only manage non-negative link-lengths.

For our purposes we are most interested in algorithms for finding the shortest paths between many nodes, that can at least manage positive link-lengths. Furthermore we are interested in efficient algorithms for re-computing shortest routes and/or their lengths after slight changes in the network. We always consider networks with nonsymmetric link-lengths: $d_{pq} \neq d_{qp}$. If the link-lengths are symmetric, fewer operations are usually necessary (when a matrix algorithm is applied exactly half the number of operations is necessary).

The graph of all shortest paths from one node to all other nodes will always be a tree (provided there is only one shortest path between a pair of nodes). For if this graph is not a tree, it means that there are other shortest paths from a node to the origin node. Even when there are other shortest paths between pairs of nodes (with the same lengths) most algorithms in fact generate a tree for the shortest paths, as will become clear in the following sections.

7.1.3   General Remarks about the Algorithms

There has been considerable development in shortest-path algorithms since Minty (1957) proposed to construct a copy of the network consisting of pieces of string with lengths proportional to the lengths of the links, then to put the origin node in one hand and the destination node in the other, to stretch and to determine the shortest route as the path with tense strings.

We are looking now for efficient algorithms that can be run on a computer. By efficient we mean:

efficient in the use of computer storage;
efficient in the use of computing time.

The necessary computer storage and computing time will be commented on in the discussion of the different algorithms and in Section 7.4. With respect to the computing time, generally speaking we must say that much depends on the particular way of coding the computer program. A well-coded program can be very much faster than a less well-coded program for the same algorithm. As an illustration, we ourselves reduced the computation time for the assignment part of the shortest path and assignment computation for the road network for the Dutch Integral Transportation Study from about 10 minutes to about 2.5 minutes just by adding one array giving the positions of some links in the network array. This importance of the coding of the program is one of the reasons that the essential part of the shortest-path computation is quite often written in the assembler language. Moreover, we will note that the computation time also depends on the structure of the network, on the actual link-lengths and even on the way in which the nodes are numbered.

There are two main types of algorithms for computing the shortest paths between many nodes:

(a) *treebuilding* algorithms;
(b) *matrix* algorithms.

In a treebuilding algorithm generally the shortest path from one node to all other nodes is found. (Using a once-through algorithm (Section 7.2.2) it is possible to compute the shortest path from one node to a subset of all nodes, just by truncation of the computation.) To get the shortest paths from n nodes to all other nodes n shortest-path trees must successively be built. In a treebuilding algorithm the structure of the network is taken into account as adequately as possible and only the connections that really exist are considered.

In a matrix algorithm the shortest paths between all nodes are obtained simultaneously. Moreover the network, the actual lengths of the shortest paths and the shortest paths themselves are stored in the form of a matrix.

The easiest way of indicating a shortest-path tree is by giving every node a *backnode*. A backnode is the preceding node in the shortest-path tree (so if we are finding shortest paths from node 1 in Figure 7.1.1 the backnode of 9 is 8, of 8 is 7 and of 7 is 1).

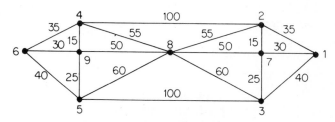

Figure 7.1.1   Network for illustrations

7.2   TREEBUILDING ALGORITHMS

The basic operation in the treebuilding algorithms is the operation of equation (7.1.1):

$$d^{aj} = \min_{(ij \,\in\, L)} (d^{ai} + d_{ij})$$

This basic operation is executed on all nodes that are 'active'. A node is called active when the basic operation must be executed on it. Different algorithms have different rules to define at each moment which nodes are active.

7.2.1   The Algorithm of Moore/Ford/Bellman

Very widely used in transportation studies is an algorithm usually called the algorithm of Moore (1957), although Ford (1956), Bellman (1958) and probably others published similar ones.

We first put all distances d^{aj} equal to infinity and we have no backnodes bn^{aj}. Then we start building the tree by making the origin active and executing the basic operation on it. During each phase of the treebuilding process which follows, every node connected with an active node is checked to see whether

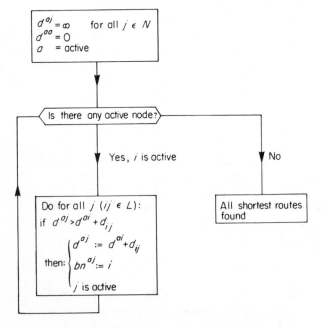

Figure 7.2.1   Flow chart of the general algorithm of
Moore/Ford/Bellman

either the original distance infinity, or the distance actually found and the route as marked by the backnode, can be replaced by a shorter one. If so, the node for which the distance has been shortened becomes active and the process is continued.

A flow chart of the general algorithm is given in Figure 7.2.1. The algorithm is stopped when there are no longer any active nodes.

First we prove that in fact all shortest paths from the origin node a are found in this way. Suppose path p $(ax_1x_2x_3 \ldots x_nj)$ has been found by the algorithm as the shortest path from a to j. In that case it holds that:

$$d^{aj} - d^{ax_n} = d_{x_nj}$$

$$d^{ax_n} - d^{ax_{n-1}} = d_{x_{n-1}x_n}$$

$$\vdots$$

$$\frac{d^{ax_1} - d^{aa} = d_{ax_1}}{d^{aj} = \sum_{il \,\in\, p} d_{il}} \qquad (7.2.1)$$

Suppose now that path p' $(ax'_1x'_2x'_3 \ldots x'_mj)$ is shorter but has not been found by the algorithm. Because there are no active nodes it must be true that:

$$d^{aj} - d^{ax'_m} \leqslant d_{x'_mj}$$

$$d^{ax'_m} - d^{ax'_{m-1}} \leqslant d_{x'_{m-1}x'_m}$$

$$\vdots$$

$$\frac{d^{ax'_1} - d^{aa} \leqslant d_{ax'_1}}{d^{aj} \leqslant \sum_{i'l' \,\in\, p'} d_{i'l'}} \qquad (7.2.2)$$

But this is contrary to the supposition that path p' is shorter than path p. So the algorithm does find the shortest routes.

An important factor for the efficiency of the algorithm is the order of succession in which the active nodes are managed. A not-too-efficient method is to give the active nodes a flag, to check the nodes every time to see if they possess a flag or not and then simply to take the first one with a flag.

Another method is to work in shifts. The first shift is node a. All nodes connected with node a form the second shift. The $(n + 1)$th shift consists of all nodes that have become active working through the nth shift. There are as many shifts as there are links in the shortest path consisting of the maximum number of links, that is a maximum of $n_N - 1$.

A third, perhaps more efficient, method is the indexing system of d'Esopo (see Pollack and Wiebenson, 1960). Node a gets the index 1 and all other nodes get index numbers in the order in which they are first reached. Then we always work with the active node with the lowest index number, so we 'stay near home'. This method bears some resemblance to the so-called 'once through'-algorithms which will be discussed in the next subsection.

It is rather difficult to give any quantitative estimations for the efficiency of the different methods. We know that each time a node is active we have to make as many additions and comparisons as there are links connected with that node. Also we know that every node is active at least once, but it can be active many times. For the road network of the Dutch Integral Transportation Study the 'shift'-method has been used. For every tree about 25,000 additions and comparisons had to be made. Because that network contained about 6,000 links, its nodes became active four times on an average.

If for some nodes the shortest distances and paths are known in advance it is profitable to use this information in the computation of the shortest-path tree. This can be done by putting at the start of the computation process the distances and backnodes of those nodes equal to their known value. This is especially relevant in the case of symmetric networks in which many shortest distances and paths may have been computed already in the shortest-path computations of nodes previously handled.

7.2.2   'Once Through' Algorithms

In the class of treebuilding algorithms, called 'once through' algorithms (the name stems from Murchland, 1967) each node is active only once and so we have to do the basic computations of the inner part of Figure 7.2.1 exactly $(n_N - 1)$ times when the whole tree is needed, while the number of additions and comparisons for the basic computations is n_L. The first 'once through' algorithms have been published by Dijkstra (1959) and Whiting and Hillier (1960), while Pollack and Wiebenson (1960) mention a personal communication with Minty which is probably the first description of a 'once through' algorithm.

During the computation process the distances of the nodes can be permanent, i.e. the shortest path for that node has been found, or tentative, i.e. a shorter path could perhaps be found. We also call the corresponding nodes permanent or tentative. At the start the origin node a gets the permanent distance zero and all other nodes get the tentative distance infinity. The basic operation (the inner part of Figure 7.2.1) is always executed on that node which was the last to become permanent. Of all tentative nodes that node with the shortest distance becomes permanent. So we have to execute the basic operation exactly $(n_N - 1)$ times.

To prove the correctness of the algorithm we can use the general proof for the algorithm of Moore/Ford/Bellman, completed with the proof that a permanent node can never become active again. The latter is easily proved by induction. Suppose that the distances to the nodes already made permanent are in fact the shortest distances to those nodes. Suppose further that node j was the last to become permanent. Then the following relationship holds: $d^{aj} \leqslant d^{*ai} + d_{ik}$ for all permanent i so the following certainty holds:

$$d^{aj} \leqslant d^{*ai} + d_{ik} + \sum_{\substack{lm \in p \\ p \text{ forms a path} \\ \text{from } k \text{ to } j}} d_{lm} \qquad (7.2.3)$$

Because all other possible paths from a to j have the form of the right hand of relationship (7.2.3) d^{aj} is in fact the real shortest distance and j can never become active again. To complete the proof it is necessary to show that the distance to the node that has become permanent first is indeed the shortest distance to that node. But this is evident because for the first permanent node n holds:

$$d^{an} = d_{an} \leqslant d_{ap}$$

so certainly:

$$d^{an} \leqslant d_{ap} + \sum_{\substack{qr \in p' \\ p' \text{ forms a path} \\ \text{from } p \text{ to } n}} d_{qr}$$

Having divided the set of nodes into a set of permanent ones and another one of tentative ones, we may make another use of this division. In the execution of the basic operation it is not necessary to test whether $d^{ai} > d^{aj} + d_{ji}$ if i is permanent, because we know that d^{ai} is smaller, since $d^{ai} = d^{*ai}$ is the shortest distance.

Because of the fact that only $(n_N - 1)$ times the basic operation need to be executed many authors state that the 'once through' algorithms are the most efficient treebuilding algorithms (Dreyfus 1969, Jansen 1970 and 1971). Though this seems indeed to be true, we must say that much depends on the efficiency of the sorting procedure used to define the next permanent node. If this is not done efficiently, many comparisons must be made to get the next permanent node; this is not necessary with the treebuilding algorithms of Section 7.2.1.

A rather efficient sorting procedure is given by Dial (1969) (see also Steenbrink, 1973) (though the procedure is called 'Moore', the algorithm of Dial is an 'once through' algorithm). Here all nodes reached are stored in an array on positions indexed by the shortest distances found. Experiments done by us showed that the algorithm of Dijkstra with a normal sorting procedure is much less efficient than the algorithm of Moore working with shifts, while the algorithm of Dial is more efficient than the algorithm of Moore working with shifts.

An important feature of the 'once through' methods is that when we are looking for the shortest routes between only a few nodes, we can stop the searching process as soon as those nodes have become permanent.

When only the shortest path between two nodes is sought, some authors propose to fan out simultaneously from both nodes. The most efficient way seems to be to add links to the two trees in an alternating way. Using a slightly changed 'once through' algorithm (two trees are built: one *from* *a* and one *to* *b*) the important thing is to decide when to stop the algorithm. Nicholson (1966) gives the right criterion.

Suppose that the distance from *a* to the node that has become permanent last, is d^a. Then we know that all nodes directly connected with the shortest path tree from node *a* have been inspected and moreover that they have a distance from node *a* greater than or equal to d^a. Suppose further that for the other tree, the distance from the node that has become permanent last to node *b*, is d^b. All nodes directly connected with that second shortest path tree have been inspected too and they have a distance to node *b* greater than or equal to d^b. It is seen easily that if the length of the shortest path from *a* to *b* actually found is smaller or equal to $d^a + d^b$ that the absolutely shortest path from *a* to *b* is found. So the stopping criterion is:

$$\text{if } d \leqslant d^a + d^b \text{ then } d \text{ is the shortest path from } a \text{ to } b$$

with:

$$d = \text{length of the shortest path from } a \text{ to } b \text{ actually found.}$$

Murchland (1967) states that when the distance for one pair of nodes is sought the saving in computation time over the normal one-ended 'once through' methods is about 50 per cent.

7.3   MATRIX ALGORITHMS

In a matrix algorithm the shortest distances and paths between all nodes are found simultaneously. We will discuss here two of the most efficient matrix algorithms namely that of Floyd and that of Dantzig. These two algorithms are for instance twice as efficient in computation time as the rather well-known cascade algorithm of Farbey, Land and Murchland (1967) for which $n_N(n_N - 2)$ additions and comparisons are necessary for the forward process and another $n_N(n_N - 1)(n_N - 2)$ additions and comparisons for the backward process.

7.3.1   The Algorithm of Floyd

The algorithm of Floyd (1962) starts with a matrix in which

$$d^{ij} = d_{ij} \quad \text{if } ij \in L$$

and:

$$d^{ij} = \infty \quad \text{if } ij \notin L.$$

We call this the initial matrix $d^{ij(0)}$. This matrix is changed in n_N steps into the final matrix $d^{ij(n_N)}$ of shortest distances for which it holds that

$$d^{ij(n_N)} = d*^{ij}$$

In the nth step we check to see if we can improve on the distance found till then by going through node n. So the basic operation is:

$$d^{ij(n)} = \min \{d^{ij(n-1)}, d^{in(n-1)} + d^{nj(n-1)}\} \qquad (7.3.1)$$

(or course the order of succession of handling the nodes is optional).

To get the shortest path itself again the backnodes are kept, only now mostly in the form of a matrix in which every row or column forms a tree. To get some understanding of the working of the algorithms we look at the way the shortest path from node 1 to node 6 in Figure 7.1.1 is found. In the 7th step the current value of d^{18} (namely $d_{12} + d_{28} = 90$) will be replaced by the correct value $d^{17} + d^{78}$. In the 8th step the current value of d^{19} will be replaced by $d^{18} + d^{89}$ and at the last step the correct value for d^{16} namely $d^{19} + d^{96}$ will be found.

The proof that all shortest paths are in fact found by the algorithm is by complete induction. First we prove that $d^{ij(1)}$ is the shortest distance from i to j if we can only use node 1 as an intermediate node:

$$d^{ij(1)} = \min \{d_{ij}, d_{i1} + d_{1j}\} \qquad (7.3.2)$$

The statement above is trivial.

Now we show that $d^{ij(n+1)}$ is the shortest distance from i to j when it is permissible to use any of the nodes from 1 to $n + 1$ as intermediate nodes if $d^{ij(n)}$ is the shortest distance from i to j when it is permissible to use any of the nodes from 1 to n as intermediate nodes. This is also trivial from:

$$d^{ij(n+1)} = \min \{d^{ij(n)}, d^{in+1\,(n)} + d^{n+1\,j(n)}\} \qquad (7.3.3)$$

for all $i \in N$, except $i = n + 1$, and all $j \in N$, except $j = i$ or $j = n + 1$.

With these two statements (7.3.2) and (7.3.3) it is proved that $d^{ij(n_N)}$ is the shortest distance when it is permissible to use all nodes as intermediate nodes, so $d^{ij(n_N)}$ is the absolutely shortest distance. The total number of additions and comparisons is $n_N(n_N - 1)(n_N - 2)$, because there are n_N matrices for which for all i's except $i = n$ (so totally $n_N - 1$) connected with all j's except $j = 1$ and $j = n$ (so totally $n_N - 2$) must be inspected.

7.3.2 The Algorithm of Dantzig

In the algorithm of Dantzig (1966) n_N matrices are also generated in which again in the nth matrix all shortest paths are present for which it is permissible to use any of the nodes from 1 to n as intermediate nodes. Besides that the nth matrix is of the order $n \times n$. So we start with one element d^{11} and get matrices of higher and higher order until the finishing complete matrix of $n_N \times n_N$ has been reached.

The operations of the nth step are as follows:

$$d^{in(n)} = \min_{1 \leqslant j \leqslant n-1} \{d^{ij(n-1)} + d_{jn}\} \quad \text{for all } 1 \leqslant i \leqslant n-1$$

$$d^{ni(n)} = \min_{1 \leqslant j \leqslant n-1} \{d_{nj} + d^{ji(n-1)}\} \quad \text{for all } 1 \leqslant i \leqslant n-1$$

and after that:

$$d^{ij(n)} = \min \{d^{ij(n-1)}, d^{in(n)} + d^{nj(n)}\} \tag{7.3.4}$$

$$\text{for all } 1 \leqslant i \leqslant n-1 \text{ and for all } 1 \leqslant j \neq i \leqslant n-1$$

The proof that the algorithm gives the correct shortest distances is again by complete induction. It is evident that for $n = 3$ (the first general case) the shortest paths found are correct:

$$d^{13(3)} = \min \{d^{11(2)} + d_{13}, d^{12(2)} + d_{23}\}$$

$$d^{23(3)} = \min \{d^{21(2)} + d_{13}, d^{22(2)} + d_{23}\}$$

$$d^{31(3)} = \min \{d_{31} + d^{11(2)}, d_{32} + d^{21(2)}\}$$

$$d^{32(3)} = \min \{d_{31} + d^{12(2)}, d_{32} + d^{22(2)}\}$$

$$d^{12(3)} = \min \{d^{12(2)}, d^{13(3)} + d^{32(3)}\}$$

$$d^{21(3)} = \min \{d^{21(2)}, d^{23(3)} + d^{31(3)}\}$$

Also it is evident from the formulation of the computations at the nth step that the nth step gives a $n \times n$ matrix of shortest distances when it is permissible to use any of the nodes from 1 to n as intermediate nodes if the $(n-1)$th step gives a $(n-1) \times (n-1)$ matrix with the same properties for $(n-1)$ nodes. So it has now been proved that the n_Nth step gives a $n_N \times n_N$ matrix of shortest distances in which it is permissible to use any of all nodes as intermediate nodes; thus they are the real shortest distances.

It is a little more difficult to obtain the total number of additions and comparisons. For the first evaluation, the evaluation of (d^{in}), $(n-1)(n-1)$ additions and comparisons are necessary for the nth matrix; for the second evaluation (d^{ni}) another $(n-1)(n-1)$ additions and comparisons are needed and finally $(n-1)(n-2)$ additions and comparisons are needed for the evaluation of d^{ij}. So the total number of additions and comparisons is:

$$\sum_{n=1}^{n_N} \{2(n-1)(n-1) + (n-1)(n-2)\} = n_N(n_N - 1)^2$$

So the algorithms of Dantzig and Floyd possess the same efficiency.

7.4 EVALUATION OF THE TREEBUILDING *VERSUS* THE MATRIX ALGORITHMS

7.4.1 Evaluation with Respect to the Necessary Computer Storage

It has been said already that a shortest-path tree is kept in storage by giving every node a backnode. For the tree, therefore, n_N elements need to be stored. Next, n_N elements are necessary for the actual lengths of the shortest paths from all nodes to the origin node. Furthermore a 'flag' is needed to indicate if we must inspect a node or not; this means another n_N positions in the storage, but it is also possible to use another much smaller array containing the nodes that must be inspected. Finally, at least $n_N + 2n_L$ positions are needed to store the network itself and the lengths of the links Altogether this means about $4n_N + 2n_L$ positions in the storage for one tree and the network. It is usually only necessary to keep one tree in the storage at the same time.

In a matrix algorithm the actual lengths of the shortest paths and the backnodes are stored in two matrices of $n_N \times n_N$ elements. It is possible to have the network and the link lengths in table form ($n_N + 2n_L$ elements) or in matrix form ($n_N \times n_N$ elements). But it is not necessary to keep this last matrix in the storage on its own, for it can usually be used as a starting tableau for the computations of the matrix algorithm. Still the necessary computer storage for matrix algorithms ($2n_N^2$ at a minimum) is much larger than for treebuilding algorithms ($4n_N + 2n_L$ at a maximum), at least when the matrix is not completely full, as will seldom be the case for transport networks.

Assuming $n_L = 4n_N$ to be generally a fairly good approximation of the number of links in a transport network, we see that the necessary computer storage for a treebuilding algorithm is about $12\,n_N$ elements and for a matrix algorithm $2n_N^2$ at a minimum. So for large networks a treebuilding algorithm is much more efficient with respect to computer storage than a matrix algorithm. This fact is one of the main reasons for matrix algorithms being far less frequently used in transport studies than treebuilding algorithms.

7.4.2 Evaluation with Respect to the Necessary Computation Time

Matrix algorithms have the advantage that they are generally easily coded into a computer program. Moreover the efficiency in computing time is exactly determinable and independent of the structure of the network. It is rather difficult to compare exactly the efficiency in computation time between treebuilding and matrix algorithms.

When the shortest paths between all nodes are sought and when all nodes are directly connected with all other nodes the matrix algorithms are generally more efficient than the treebuilding algorithms. But when the nodes are only directly connected with a few other nodes (say four again) and when only the shortest paths between a subset of the nodes are sought (for instance the origins and destinations), then the treebuilding algorithms are more efficient with respect to computation time. So, in general treebuilding algorithms seem to be much more useful than matrix algorithms for transport networks.

7.5   HEURISTIC PROCEDURES

As for all complex and much computing-time-consuming problems, so for the shortest-path-finding problem heuristic procedures have been proposed. For sometimes one may be content with a reasonably short path and the absolutely shortest path is not needed, especially when it would take much more computation time to find it.

7.5.1   A Heuristic Procedure for the Shortest Path between Two Nodes

Mills (1968) proposes a heuristic treebuilding algorithm for finding the shortest route from a to b. In this algorithm less promising routes are not inspected further. To estimate if a route is promising every 'reached' node i gets not only a distance d^{ai} but also a 'value', composed of d^{ai} and an estimation for d^{ib}. For this last estimation we can take, for instance, the distance as the crow flies. During the course of the algorithm we always take that active node with the lowest 'value' as the node from which branching is carried out. When node b is reached we check whether there are any nodes with 'values' lower than the d^{ab} bound. If so, we continue the computation from the node with the lowest 'value'. If not, the computation is stopped. Of course we do not know for sure that the shortest route is found. Note that the 'value' for node i is *not* a lower bound for the length of the path from a to b through i, for it is not known if d^{ai} is permanent or not. However, the route found seems to be a reasonably short one, if not the shortest one.

7.5.2   Heuristic Procedures for the Shortest Paths between Many or All Nodes

Jansen (1971) makes the remark that it seems plausible that when searching for the shortest paths between many nodes, it should be possible to turn to advantage the fact that some nodes are geographically very close together. We tried to work out this thought for the use of treebuilding algorithms in very large networks.

It seems plausible that two shortest-path trees originating from nodes close together are identical at some 'distance' from the originating nodes. Even in the very small network of Figure 7.1.1 the trees from the nodes 1 and 2 are identical left of the line 483. So it should be possible to use the tree from one node for the other node too, at some 'distance' from the two originating nodes. The important thing is the procedure of checking whether the same trees are found.

Unfortunately, though, two trees can be identical for some nodes without necessarily being so for all nodes. Even if the two trees are identical on a certain ring around the originating nodes they are not necessarily identical outside this ring. Interpreted geometrically, we see that the shortest paths are different for nodes in the neighbourhood of the central perpendicular of the two origins, even at a considerable distance from the origins. So it does not seem to be possible to use the trees of neighbouring nodes as directly as proposed above. It may be possible, however, to use some parts of the trees of neighbouring nodes and to make a considerable saving in computation time in that way.

One may look at this kind of procedure as a sort of node aggregation. Instead of computing all trees for the different nodes of one area, we compute one tree for some supernode leaving the routes for the nodes inside the area itself to some other computation.

Another procedure for simplifying the shortest-path computations has been proposed by Tanner and others (1972). They use a simplified representation of the network. They work with idealized networks consisting of concentric circles and equally spaced radials. In this way many nodes are symmetrically located and so the shortest-path trees for these nodes possess the same structure. This feature makes it possible to make far fewer computations to get all shortest-path trees than in the common situation.

Though this procedure is not a heuristic one in the real sense of the word, it may be of interest to present it here to show how efforts are made to (over)simplify the real-life network, with the idea of getting simple computations, in the hope that this simple model will give a reasonable description of reality.

7.6 PARTITIONING

The reduction of computation time and computer storage for computations with large matrices by means of partitioning principles is very well-known. Partitioning is also useful in the matrix computations for the finding of the shortest routes. When it is not so that every node is connected with every other node, so that there are, say, $4n_N$ links instead of $n_N(n_N - 1)$ (a very common occurrence in transport networks), and especially when it is possible to divide the original set of nodes into a number of subsets loosely connected with each other, a partitioning will be very profitable. We will here discuss a partitioning procedure for the algorithm of Floyd, proposed by Hu (1968).

First we divide the set of nodes N into three subsets N^A, B^B and N^Z with the properties that N^A and N^B are totally disconnected and N^Z is the subset of all nodes that connect N^A with N^B so:

$$d_{ij} = \infty \qquad i \in N^A \text{ and } j \in N^B \text{ or } i \in N^B \text{ and } j \in N^A$$

and

$$N^Z = \{j/d_{ij} \neq \infty \text{ and } d_{jk} \neq \infty, \ i \in N^A, \ k \in N^B \text{ or } d_{kj} \neq \infty$$
$$\text{and } d_{ji} \neq \infty, i \in N^A, k \in N^B\}$$

We define the conditional shortest path $d^{ab}(N^P)$ as the shortest path from a to b for which all nodes of the path are contained in the set N^P. The algorithm of Hu requires four steps:

Step 1:

compute $d^{ab}(N^A \cup N^Z)$ \qquad for all $a \in N^A \cup N^Z$ and $b \in N^A \cup N^Z$;

Step 2:

using the results of step 1 for $a \in N^Z$ and $b \in N^Z$ compute d^{ab} for all $a \in N^Z \cup N^B$ and $b \in N^Z \cup N^B$;

we obtain then:

$d^{ab}(N^A \cup N^Z \cup N^B) = d*^{ab}$ \quad for all $a \in N^Z \cup N^B$ \quad $b \in N^Z \cup N^B$;

Step 3:

using the results of step 2 for $a \in N^Z$ and $b \in N^Z$ compute d^{ab} for all $a \in N^A \cup N^Z$ and $b \in N^A \cup N^Z$;

we obtain then:

$d^{ab}(N^A \cup N^Z \cup N^B) = d*^{ab}$ \qquad for all $a \in N^A \cup N^Z$ and $b \in N^A \cup N^Z$;

Step 4:

using the results of step 2 and step 3 compute the remaining shortest paths for $a \in N^A$ and $b \in N^B$ or $a \in N^B$ and $b \in N^A$ by

$$d^{ab} = \min_{i \in N^Z} \{d^{ai} + d^{ib}\} \qquad \text{for all } a \in N^A \text{ and } b \in N^B$$

and

$$d^{ab} = \min_{i \in N^Z} \{d^{ai} + d^{ib}\} \qquad \text{for all } a \in N^B \text{ and } b \in N^A.$$

For steps 1, 2 and 3 we use a matrix algorithm for matrices of the order $(n_A + n_X) \times (n_A + n_X)$, $(n_X + n_B) \times (n_X + n_B)$ and $(n_A + n_X) \times (n_A + n_X)$ respectively (with n_A the number of nodes of subset N^A and so on). To prove that the partitioning gives the correct shortest paths we prove the steps 2, 3 and 4. Then we have proved that all shortest paths have been found.

The proof of step 2 is as follows: if the shortest route from a to b is totally contained in the $N^Z \cup N^B$, it is trivial that the result is the real shortest path.

If not we can decompose the shortest path into paths that are totally contained in $N^Z \cup B^B$ or in $N^A \cup N^Z$. So we get for instance:

$$d^{*ab} = d^{*ai_1} + d^{*i_1i_2} + d^{*i_2i_3} + d^{*i_3i_4}$$
$$\begin{matrix} a \in N^B & i_1 \in N^Z & i_2 \in N^A & i_3 \in N^A \\ i_1 \in N^Z & i_2 \in N^A & i_3 \in N^A & i_4 \in N^Z \end{matrix}$$
$$+ \cdots + d^{i^n}$$
$$\begin{matrix} i^n \in N^Z \\ b \in N^B \end{matrix}$$

for all these elements the shortest path is found in the first or second step, so the correct shortest distance d^{*ab} is found. If the decomposition of the path should be otherwise, the conclusion about the correct shortest distance would be the same.

Step 3 is proved in a similar way to step 2. Step 4 is proved very simply: because N^A and N^B are disconnected the shortest path from a node of N^A to a node of N^B or in reverse must go through a node of N^Z. So $d^{*ab} = d^{*ai} + d^{*ib}$ with $i \in N^Z$. It has been proved above that the correct values for d^{*ai} and d^{*ib} are found in the preceding steps. So inspecting all possibilities for the intermediate node i and taking that node i which gives the minimum value for $d^{*ai} + d^{*ib}$ yields the correct solution.

When we take for simplicity n^3 additions and comparisons for a shortest route computation for a matrix of $n \times n$ (instead of $n(n-1)(n-2)$) the total number of additions and comparisons is:

$(n_A + n_X)^3$ for step 1

$(n_X + n_B)^3$ for step 2

$(n_A + n_X)^3$ for step 3

$2n_A n_B n_X$ for step 4

so, totally, $2(n_A + n_X)^3 + (n_X + n_B)^3 + 2n_A n_B n_X$.

Using a normal matrix algorithm $(n_A + b_B + n_X)^3$ additions and comparisons are necessary, which is generally much more, especially when n_X is small compared to n_A and n_B. For instance with $n_A = 0.4n_N$, $n_B = 0.5n_N$ and $n_X = 0.1n_N$, the partitioning algorithm needs $0.486\,n_N^3$ additions and comparisons against n_N^3 for the unpartitioned one. Hu shows that it is also possible to partition the matrix into more submatrices, perform the steps 1 and 2 going along the diagonal from up left to down right, perform step 3 going in the reverse direction and finish with step 4 for all remaining parts of the matrix. It is, of course, also possible to partition a submatrix and so on.

7.7 SMALL CHANGES IN THE NETWORK

In evaluation and optimization procedures it is a very common situation that, after having computed the shortest distances and shortest paths for some network, we must recompute the shortest paths and/or distances for a slightly different network. We have met this situation, particularly in the discussion of branch and bound methods in Section 4.3. In such a case it is not always necessary to repeat the computation completely. Instead we can do the job much quicker using the information obtained from the computation for the original network. As an illustration we will discuss here two very similar algorithms that recompute the shortest distances and/or paths after a change in the length of one link (which includes the addition or deletion of a link). First we observe that when the shortened link pq does not form the shortest route from p to q the shortening of the link has no effect at all.

7.7.1 Recomputing All Shortest Paths after a Change in the Length of One Link according to Loubal

Loubal (1967) proposes an algorithm to recompute all shortest paths using the formulation of treebuilding algorithms; but the principle can also be used in matrix algorithms.

We suppose that the length of link pq will be changed from d_{pq} to d'_{pq}. The problem is to find all $d^{*\prime\,ab}$ and all corresponding routes when all $d^{*\,ab}$ with their routes are known. We further suppose that it is known in advance that the length of link pq will be changed. During the computation of the original shortest routes we compute also 'the shortest routes from and to link pq', that means we connect p and q by links with zero length with an artificial node r and compute also the shortest routes from and to r. Furthermore we label all nodes to indicate if the shortest routes from and to r go through p or through q. So we divide the original set of nodes N into two mutually exclusive and totally exhaustive subsets N^p and N^q ($N^p \cup N^q = N$ and $N^p \cap N^q = \phi$).

We assume here that the shortest routes from a to r and from r to a both go through p or q. This may not always be true. In the case where the two paths go through different nodes the evaluation statements (7.7.1) to (7.7.5) must be changed a little. It is easy to write down the correct evaluation statements for that case (using subsets N'^p, N''^p, N'^q and N''^q).

Now we come to the problem of recomputing the shortest routes and/or distances for all relations ab when the length of link pq is shortened. To do that we check first to which subset of nodes the nodes a and b belong. If they belong to the same subset the shortening of the length of the link pq cannot have any effect on the shortest distance and route. Nor when $a \in N^q$ and $b \in N^p$ can a new shortest route be found.

Only in the case where $a \in N^p$ and $b \in N^q$ may the link-shortening have any effect. In that case a simple evaluation must be made, comparing the original distance with the distance obtained using the shortened link pq.

We give the statements fully below:

(a) if $a \in N^p$ and $b \in N^p$ then $d^{*\prime ab} = d^{*ab}$ (7.7.1)

Proof:

We assume here that the shortened link pq still has a positive length ($d'_{pq} > 0$), however small it may be.

$$d^{*pb} \leqslant d^{*qb}$$

$$d^{*ap} + d^{*pb} \leqslant d^{*ap} + d^{*qb}$$

$$d^{*ap} + d^{*pb} < d^{*ap} + d'_{pq} + d^{*qb}$$

$$d^{*\prime ab} \leqslant d^{*ap} + d^{*pb} < d^{*ap} + d'_{pq} + d^{*qb}$$

so pq cannot be obtained in the shortest route from a to b and $d^{*\prime ab} = d^{*ab}$;

(b) if $a \in N^q$ and $b \in N^q$ then $d^{*\prime ab} = d^{*ab}$

Proof:

$$d^{*aq} \leqslant d^{*ap}$$

$$d^{*aq} + d^{*qb} \leqslant d^{*ap} + d^{*qb}$$

$$d^{*aq} + d^{*qb} < d^{*ap} + d'_{pq} + d^{*qb}$$

$$d^{*\prime ab} \leqslant d^{*aq} + d^{*qb} < d^{*ap} + d'_{pq} + d^{*qb}$$

so pq cannot be obtained in the shortest route from a to b and $d^{*\prime ab} = d^{*ab}$;

(c) if $a \in N^q$ and $b \in N^p$ then $d^{*\prime ab} = d^{*ab}$ (7.7.2)

Proof:

$$d^{*aq} \leqslant d^{*ap}$$

$$d^{*aq} + d^{*qb} \leqslant d^{*ap} + d^{*qb}$$

and so on (see (b));

(d) if $a \in N^p$ and $b \in N^q$ then $d^{*\prime ab} = \min \{d^{*ab}, d^{*ap} + d'_{pq} + d^{*qb}\}$ (7.7.4)

Proof:

d^{*ab} is the shortest route when link pq has length d_{pq};

only a path using the shortened link pq can be shorter, so we have only to compare the two possibilities mentioned above.

It is obvious that the use of the algorithm of Loubal is much quicker than the repeated use of a 'normal' algorithm to compute the shortest routes and distances (see also our remarks about the efficiency of the algorithm of Murchland (1969) made in Section 7.7.2).

Loubal adds the remark that it is not even necessary to evaluate all ab with a $a \in N^p$ and $b \in N^q$, since, if for some node a' or b' no shorter path can be found, it will also be impossible for all nodes on the branches 'behind' them. This idea may be put into practice by starting to inspect the relations aq or bp. If $d^{*\,aq} \leqslant d^{*\,ap} + d'_{pq}$ no more relations from a need to be inspected and if $d^{*\,pb} \leqslant d'_{pq} + d^{*\,qb}$ the same is true for the relations to b (for if $d^{*\,aq} \leqslant d^{*\,ap} + d'_{pq}$ or if $d^{*\,pb} \leqslant d'_{pq} + d^{*\,qb}$ certainly $d^{*\,ab} \leqslant d^{*\,ap} + d'_{pq} + d^{*\,qb}$). In this form the idea has also been put forward by Murchland (1970).

It is also easy to update the backnodes, as can readily be seen. The disadvantage is that it is then necessary for all trees, including the tree from the artificial node r, to be available.

When a link is shortened for both directions (as will be generally the case) that is to say when:

$$d'_{pq} < d_{pq} \quad \text{and} \quad d'_{qp} < d_{qp}$$

the same algorithm can be used.

The statements (7.7.1) and (7.7.2) are identical and the statements (7.7.3) and (7.7.4) change into:

If $a \in N^q$ and $b \in N^p$ or $a \in N^p$ and $b \in N^q$ then:

$$d^{*'\,ab} = \min \{d^{*\,ab}, d^{*\,ap} + d'_{pq} + d^{*\,qb}, d^{*\,aq} + d'_{qp} + d^{*\,pb}\} \qquad (7.7.5)$$

This is the original formulation of the algorithm of Loubal.

It is also possible to use the algorithm for situations where more links are shortened at the same time. In such situations one can quite simply obtain relationships similar to those found with one-link shortening. The only difficulty is that we must inspect all possible combinations of original shortest paths and shortened links. This means that when k links are included in the shortest path all $k!$ possible orders of succession must be inspected. Moreover when n links are shortened $n, n - 1, n - 2, \ldots, 0$ links can be contained in the shortest path and when k links are obtained in the shortest path all of the $\binom{n}{k}$ combinations are possible. So, recapitulating, the total number of possibilities with n links shortened is:

$$\sum_{k=0}^{n} k! \binom{n}{k}$$

For $n = 6$ this number is already 1,238 and the number increasing very rapidly with increasing n.

When a certain link is lengthened (or when a link is deleted from the network) a similar procedure can be used. For only the shortest paths for the relations ab with $a \in N^p$ and $b \in N^q$ can be affected; so only these relations need to be inspected. Moreover it may be useful to start by testing whether the link pq belongs to the original shortest path tree (by testing if

$d*^{aq} = d*^{ap} + d_{pq}$ or if $d*^{pb} = d_{pq} + d*^{qb}$). If this is not the case, the shortest paths for ab will again be unaffected. Unfortunately, for the inspection of the possibly-affected shortest distances no evaluation procedure as simple as statement (7.7.4) is available. So there is the possibility that many shortest path trees will have to be totally rebuilt.

Halder (1970) proposes the 'method of competing links'. There exists a set of links L^D the removal of which would disconnect every node of N^p from every node of N^q. So a shortest path from a node of N^p to a node of N^q will always contain one of the (competing) links of L^D. So only the following inspection is necessary (the new lengthened link pq being included in the set L^D):

$$d*'^{ab} = \min_{rs \in L^D} (d*^{ar} + d_{rs} + d*^{sb}) \qquad (7.7.6)$$

However, if possible, it is much more profitable first to compute the shortest paths for the new network in which the related link has its longest length and next to apply the recomputing procedure to the original network.

7.7.2 Recomputing All Shortest Paths after a Change in the Length of One Link according to Murchland

Murchland (1969) proposes an algorithm formulated as a matrix algorithm but the principle can also be used for treebuilding algorithms. Again we assume that all $d*^{ab}$ and the corresponding routes are known, that the length of link pq is shortened from d_{pq} into d'_{pq}, (that $d*'^{pq} = d'_{pq}$) and that the problem is to find the new distances $d*'^{ab}$ and their corresponding routes. Murchland gives the following relationships:

$$d*'^{aq} = \min \{d*^{aq}, d*^{ap} + d'_{pq}\} \quad \text{for all } a \in N, a \neq q$$

and:

$$d*'^{ab} = \min \{d*^{ab}, d*'^{aq} + d*^{qb}\} \quad \text{for all } a \in N \text{ and all } b \in N,$$

$$b \neq a, \quad b \neq q \qquad (7.7.7)$$

It can be easily proved that these evaluations give the correct solution for all $d*'^{ab}$. It is obvious that the updating of the routes is not difficult either. The total number of additions and comparisons is $(n_N - 1) + n_N(n_N - 2) \approx n_N^2$. Compared with the total number of necessary additions and comparisons for the matrix algorithms of Floyd and Dantzig of $n_N(n_N - 1)(n_N - 2) \approx n_N^3$ one can see how much saving in computation time has been made. On the other hand it is possible to speed up the original algorithm of Murchland (Murchland, 1970) for:

$$\text{if } d*'^{aq} = d*^{aq} \text{ then } d*'^{ab} = d*^{ab}. \qquad (7.7.8)$$

It is easy here to derive the algorithm for the case where the link is shortened for both directions:

$$d^{*\prime\,aq} = \min\{d^{*\,aq}, d^{*\,ap} + d'_{pq}\} \quad \text{for all } a \in N, a \neq q$$

$$d^{*\prime\,ap} = \min\{d^{*\,ap}, d^{*\,aq} + d'_{qp}\} \quad \text{for all } a \in N, a \neq p$$

$$d^{*\prime\,qa} = \min\{d^{*\,qa}, d'_{qp} + d^{*\,pa}\} \quad \text{for all } a \in N, a \neq q$$

$$d^{*\prime\,pa} = \min\{d^{*\,pa}, d'_{pq} + d^{*\,qa}\} \quad \text{for all } a \in N, a \neq p$$

and

$$d^{*\prime\,ab} = \min\{d^{*\,ab}, d^{*\prime\,aq} + d^{*\prime\,qb}, d^{*\prime\,ap} + d^{*\prime\,pb}\} \quad \text{for all } a \in N \text{ and}$$

$$\text{all } b \in N, a \neq b, a \neq p, a \neq q, b \neq p, b \neq q \qquad (7.7.9)$$

This is the original formulation of Murchland (1969). Similar speeding-up measures to those discussed for the one-way-shortened link can be taken here. Of course, it is also possible to use a similar but more extensive algorithm for the lengthening of a link (see Section 7.7.1).

REFERENCES

Bellman, R. E. (1958). On a routing problem. *Quarterly of Applied Mathematics*, **16**, 87–90.

Clercq, F. le (1972). A public transport assignment method. *Traffic Engineering and Control* (June); also in Dutch (1972): *Verkeerstechniek*, **23**, No. 6 (June).

Dantzig, G. B. (1966). *All Shortest Routes in a Graph*, Technical Report 66–3, Operations Research House, Stanford University.

Dreyfus, S. E. (1969). An appraisal of some shortest path algorithms, *Operations Research*, **17**, No. 3 (May–June).

Dial, R. (1969). Algorithm 360, shortest path forest with topological ordering. *Communications of the Association of Computing Machinery*, **12**, 632.

Dijkstra, E. W. (1959). A note on two problems in connexion with graphs. *Numerische Mathematik*, **1**, 269–271.

Farbey, B., Land, A. H., and Murchland, J. D. (1967). The cascade algorithm for finding all shortest distances in a directed graph. *Management Science*, **14**, (September).

Floyd, R. W. (1962). Algorithm 97, shortest path. *Communications of the Association of Computing Machinery*, **5**, 345.

Ford, L. R. (1956). *Network Flow Theory*, The Rand Corporation, P-923.

Hamerslag, R., in collaboration with Steenbrink, P. A. (1970). Het voorspellen van vervoersstromen. Paragraaf 5. Reisroute, reistijden en trajektbelasting. *Verkeerstechniek*, **21**, No. 7 (July).

Halder, A. K. (1970). The method of competing links. *Transportation Science*, **4**, 36.

Hoffman, W. and Pavley, R. (1959). A method for the solution of the *n*th best path problem. *Journal of the Association of Computing Machinery*, **6**, 506–514.

Hu, T. C. (1968). A decomposition algorithm for shortest paths in a network. *Operations Research*, **16**, 91–102.

Jansen, G. R. M. (1970). *Het Zoeken van een Kortste Route in een Netwerk; een Kwalitatieve Analyse van Enkele Algorithmes*. Research Report 70–1, Transportation Research Laboratory, Delft University of Technology, Delft.

Jansen, G. R. M. (1971). Het zoeken van een kortste route in het netwerk. *Verkeerstechniek*, **22**, No. 4 (April).

Kirby, R. F., and Potts, R. B. (1969). The minimum route problem for networks with turn penalties and prohibitions. *Transportation Research*, **3**, 397–408.

Loubal, P. S. (1967). A network evaluation procedure, *Highway Research Record*, 205.

Mills, G. (1968). A heuristic approach to some shortest route problems. *Canadian Operational Research Society Journal* (March).

Minty, G. (1957). A comment on the shortest route problem. *Operations Research*. **5**, 724.

Moore, E. F. (1959). The shortest path through a maze. *Proceedings of an International Symposium on the Theory of Switching, Part II, April 2–5, 1957*, The Annals of the Computation Laboratory of Harvard University 30, Harvard University Press, Cambridge Mass.

Murchland, J. D. (1967). *The 'Once Through' Method of Finding All Shortest Distances in a Graph from a Single Origin*, Transport Network Theory Unit Report LBS-TNT 56, London Graduate School of Business Studies, London.

Murchland, J. D. (1969). *A Fixed Matrix Method for all Shortest Distances in a Directed Graph and for the Inverse Problem*. Transport Network Theory Unit Report LBS–TNT 91, London Business School, London.

Murchland, J. D. (1970). *A Fixed Method for All Shortest Distances in a Directed Graph and for the Inverse Problem*, Ph.D Thesis, University of Karlsruhe.

Nicholson, T. A. J. (1969). Finding the shortest route between two points in a network, *Computer Journal*, **9**, 275.

Pollack, M., and Wiebenson, W. (1960). Solution of the shortest route problem: A review. *Operations Research*, **8**, 224.

Ribbeck, K. F. (1971a). Comparison of some optimal route choosing algorithms. *Proceedings of the Planning & Transport Research & Computation Conference on 'Urban Traffic Models Research'*, London.

Ribbeck, K. (1971b). *Routen Suchen in Stadtstrassennetzen*, Ph.D. Thesis, Rheinisch–Westfalische Technische Hochschule Aachen.

Steenbrink, P. A. (1973). Treebuilder-algoritmen voor het vinden van de kortste route in een netwerk, *Verkeerstechniek*, **24**, No. 4 (April).

Tanner, J. C., Gyenes, L., Lynam, D. A. and Magee, S. V. (1972). Christal: A strategic model for urban transport planning. Paper presented at *The Planning & Transport Research & Computation Symposium on 'Urban Traffic Models Research'*, Integrated Models Seminar, London (May).

Whiting, P. D., and Hillier, J. A. (1960). A method for finding the shortest route through a road network. *Operational Research Quarterly*, **11**, No. 1/2.

PART II

A Case Study: The Optimization of the Dutch Road Network

A Case Study: The Optimization of the Dutch Road Network

The second part of this book gives a description of a case study: the optimization of the Dutch road network. The study forms a part of the Dutch Integral Transportation Study undertaken by the Netherlands Economic Institute at the behest of the Minister of Transport. The aim of this study was:

(a) To forecast the total future transport in the Netherlands (by road, by rail, by barge, by pipeline and, if relevant, by air) for the years 1980, 1990 and 2000.

(b) To define with respect to this the total need for interurban highways (particularly length, number of lanes and so on) and the need for railways for the years 1980, 1990 and 2000, on the basis of a forecast of the direction and size of the transport flows between the different transport zones.

(c) To define the amount of the investments necessary for (b).

The study was managed by Professor L. H. Klaassen and J. A. Bourdrez. The results have been reported in 1972 (Nederlands Economisch Instituut, 1972) in a main report and seven annexes. Annex V gives a full description of the method used and the results of the optimization of the road network.

The method of the stepwise assignment according to the least marginal objective function described in Chapter 5 of this book was used for the optimization of the road network. To this end a set of computer programs was developed by a group of systems-analysts and programmers of the Research and Planning Department of the Netherlands Railways, under the management of P. A. Steenbrink. J. Perton was in charge of the computer coding and J. A. Kant and M. Koss also worked on the project.

In Chapter 8 of this book we will point out the position of the optimization of the road network in the whole of the transport study and the defining of the optimal spatial structure. Chapter 9 gives a description of the objective function used in the Dutch Integral Transportation Study. Chapter 10 deals with the subproblem of the transport network optimization problem: the optimal number of lanes for a given traffic flow. Finally in Chapter 11 the application of the stepwise assignment according to the least marginal objective function is described. This is done on the basis of the results of a test computation of which more material is available than of the final computations. See annex V of the final report of the Dutch Integral Transportation Study for an ample description of the results of the definitive computations.

REFERENCE

Nederlands Economisch Instituut (1972). *Integrale Verkeers- en Vervoerstudie*, Staats-uitgeverij, 's-Gravenhage. (An abridged English version will be provided by the Dutch Ministry of Transport, The Hague, 1973.)

8

The Position of the Optimization of the Road Network in the Whole Transport Study and in Defining the Optimal Spatial Structure

The close relationship between the spatial structure and the transport infrastructure has been pointed out already in Section 3.1. It is obvious that the optimal solution for this whole system must be defined. However it is also obvious that this is no simple affair; defining the objective function especially is very difficult. Therefore we turned in Section 3.1 to the optimization of the transport infrastructure for a given spatial structure. In the Dutch Integral Transportation Study the spatial structure is also considered to be a datum (see annex II of the final report of this study for a description of this given spatial structure).

The problem is then stated as the optimization of the transport system for a given spatial structure with the infrastructure as decision variable, as discussed in Section 3.2. Some possibilities for the objective function have been discussed in Section 3.2.2. There we spoke about the 'costs' and the 'benefits' of the transport system for the society. The maximization of the positive difference between these benefits and costs is an important possibility for the objective.

However, defining the social benefits caused by the trips is especially difficult. This question is closely related to the social evaluation of the spatial structure. To avoid these difficulties, in the Dutch Integral Transportation Study the social benefits are considered to be constant; that is, it has been carried out with constant trip-matrices. The statement of the problem is reduced, then, to the minimization of the total social costs. This possibility for the objective has also been mentioned in Section 3.2.2.1.

There the costs were divided into three categories:

(a) costs directly related to the trip and generally borne by the trip-makers (users' costs);

(b) costs related to the construction and maintenance of the infrastructure networks (generally borne by the regulating authorities);

(c) costs related to the infrastructure or the use of it, but generally borne by 'third parties', so-called external costs.

Defining society as the summation of all private ('tripmakers' and 'third parties') and public households, the summation of the costs mentioned above can be defined as the total social costs of the transportation system.

The different cost factors taken into the objective function as related to car traffic will be discussed in Chapter 9. The costs related to public transport have been treated in annex VI of the final report of the Dutch Integral Transportation Study. The total costs consist, among other things, of travel time costs, vehicle operating costs, accident costs, investments and maintenance. The costs to the environment have not been included in the objective function, but they did play a role when the constraints were defined. Among the factors considered in defining possible places for road construction and restraints to be attached to the different links was that of possible damage to the environment.

The Dutch Integral Transportation Study has been concerned with the interurban infrastructure. Therefore only the costs of and on the interurban infrastructure have been included in the objective function.

The trip-matrices used are matrices for the traffic on workdays. So only the users' costs incurred on workdays are included in the objective function.

The optimization of the infrastructure is a dynamic problem and it is necessary to minimize the social costs for the whole period from 1970 to 2000 and after (compare Section 6.1). However, on account of the computational complexity of this problem, a simpler system has been used. The trip-matrices have been defined for three years, namely 1980, 1990 and 2000. For all three years the matrices are supposed to stay constant after the related year and those networks are defined which will minimize the total costs (discounted to the present value at the starting year) for the whole period (to infinity in principle) starting at, respectively, 1980, 1990 and 2000. On the basis of the resulting networks a coherent investment scheme for the period 1970–2000 can be defined, for which it is not, of course, necessary to assume constant trip-matrices. Furthermore other factors to be included in the objective function can be considered while defining the investment scheme.

An optimal transport infrastructure means a coherent system of rail and road networks (compare Section 6.2). However for a correct optimization it is always necessary for the objective function and the decision variables to fit well together. If a certain value for a decision variable has an effect that is expressed in the objective function for one part but not for the other part, when this second part is of great importance, a precise optimization of that objective function makes little sense. This effect is seen in the joint optimization of rail and road networks. For it is in this case when a choice has to be made between a road and a rail connection, that many factors not included in the objective function play a role. We mention among other things ιe costs in the towns, users' costs also considered as investments, and nvironmental costs.

As a first approximation for the joint optimization the following system as been used. For every year the model for the modal split is applied with ιvo different assumptions for certain basic data (see annex III of the final ιport of the Dutch Integral Transportation Study). This results in a 'high' ιrecast for car traffic complementary with a 'low' forecast for rail transport ιnd on the other hand a 'low' forecast for car traffic and a 'high' forecast ιr rail transport. For both sets of trip-matrices the optimal networks are efined. The inspection and comparison of these two sets of networks may ιach something of the optimal coherent system of road and rail infrastruc- ιre. In this comparison not only must the objective function be considered, ιt also factors that have not been included in the objective function but are ill of importance to the decisions. For the separate optimization of road ιnd rail infrastructure the exclusion of environmental and urban costs is ιss serious. These costs depend for a large part only on the total number ι kilometres travelled by the different modes and the total number of persons ι vehicles entering and leaving the town. The division of these flows among ιe different links of the networks is of course important for these cost factors ιt may not be very important. Furthermore these factors are taken into ιnsideration at some points in the form of constraints.

For the optimization of the rail networks in the Dutch Integral Transporta- ιn Study use has been made of the method of Barbier extended by Haubrich, ι described in Section 4.4.3.2 (see annex VI of the final report of the Dutch ιtegral Transportation Study for a complete description of the method ιd parameters used). For the optimization of the road network the method ι the stepwise assignment according to the least marginal objective function described in Chapter 5 has been applied, followed by a descriptive assign- ιent to the network found and an optimal adjustment of the dimensions ι the roads to the traffic flows found in this last assignment (compare Section 4.2.1).

The application of different methods must be explained by the shape ι the cost functions for the road and rail network (see also Section 6.2). ι the number of lanes is considered a continuous variable the social costs ι a road form what is almost a convex function of the traffic flow (Sections).4 and 5.3). This is certainly true in the case of an extension of an existing ιad with four or more lanes. The assumption of continuity of the number of ιnes does not seem too serious because, in reality, there are a number of ιssibilities (2-, 4-, 6-, 8-, 10- and 12-lane roads). However, for railway ιtworks the choice is quite often between constructing or not constructing, ιtween closing down or not closing down. So a very discrete decision ιriable is involved here and it hardly seems possible to assume a continuous ιmension. Moreover the exploitation costs are generally concave instead of

convex for the relevant interval. Both affairs give rise to a clear non-conve
relationship between objective function and flow. For this reason the stepwis
assignment according to the least marginal objective function has not bee
applied for the railway network. Because of the very short computation tim
(comparable with the computation time necessary for one assignment o
the trip-matrix to the network) and the advantages over heuristic method
the stepwise assignment according to the least marginal objective functio
has been applied for the optimization of the road network.

The application of different algorithms for two similar parts of one proble
can be dangerous in the comparison of the results of both parts. In th
interpretation of the joint results for the road and rail networks this mu
be taken into account. However, when the optimal solutions are not foun
both methods fortunately tend to 'errors in the same direction'. The metho
of Barbier as well as the stepwise assignment according to the least margin
objective function generally tend to give a better minimization of the user
costs than of the investments (see Section 4.4.3 and Chapter 5; the stateme
is certainly not always true).

It has been said that the optimal road network is defined for a give
matrix of trips by car. An important question here is whether the quality o
the resulting network is such, with respect to the possibilities for the oth
transport modes too, that this trip-matrix is really realized. The consistenc
between starting-points and results is involved. While defining the spati
structure and forecasting the demand for transport, it is necessary to mak
many assumptions about the quality of the infrastructure networks. Thes
assumptions can be only verified at the end of the optimization process
for the road and rail infrastructure. If the results are not consistent with th
assumptions, it may be necessary to introduce feedbacks such that th
assumptions agree with the results. The need for and consequences of thes
feedbacks are discussed in more detail in Section 11.6.

Chapter 9 deals with the objective function applied to the road netwo
optimization for the Dutch Integral Transportation Study. The subproble
of the stepwise assignment according to the least marginal objective functio
namely the defining of the optimal number of lanes for a given traffic flo
is dealt with in Chapter 10. The master problem, the defining of the traff
flows in such a way that the objective function is minimized, is dealt with
Chapter 11. This will be done on the basis of a test computation. In th
chapter a comparison is made between the results of the stepwise assig
ment according to the least marginal objective function and the resu
obtained by an optimal adjustment of the dimensions of the roads to th
traffic flows resulting from a descriptive assignment to a network in whi
all links have their maximal dimension. Finally we pay some attentic
to the possible interpretation and evaluation of the results for the roa
network.

REFERENCE

Nederlands Economische Instituut (1972). *Integrale Verkeers- en Vervoerstudie*, (Final report and annexes), Staatsuitgeverij, 's-Gravenhage.

9

The Objective Function

9.1 GENERAL COMPOSITION OF THE OBJECTIVE FUNCTIO

9.1.1 General Remarks

In this chapter we will discuss the elements of the objective function as was stated for the Dutch Integral Transportation Study. The aim is to show fully elaborated objective function and the problems met in elaborating th objective function. We are aware that there are other ways of stating an elaborating the objective function. The different elements of the objecti function are all given in Dutch guilders, but it will be clear that any unit currency could be substituted. The minimization, by choosing the dimension of the road network, of the total social costs of transport for a given amou of car traffic is dealt with. So all cost factors that may be influenced by t total amount of car traffic but not by the special routes used and/or dime sions of the road are not relevant to the problem and are consequently n included in the objective function.

The objective function consists of the total social costs of car traffic on t road network. These are the summation of the social costs over all links ar nodes of the network. We will give the different cost factors per kilomet and sometimes qualified by certain indicators pertaining to the related lir or node. The total costs for the network are found by the multiplication of t costs per kilometre by the length of the link, and the summation over a links and nodes. The users' costs on the nodes are assumed to be included the users' costs on the links and are so not explicitly taken into accour The investments are taken both for links and nodes. All costs are given as function of the dimension or the dimension and the traffic flow.

The dimension of a road or the number of lanes is expressed sometimes the number of passenger car units (p.c.u.) that can pass a point of the roa during an hour. In that case one lane of a highway is expressed in 2,000 p.c. per hour. (This does not say anything however about the desired or maximal possible number of p.c.u. per lane passing a point of a road per hour.) T traffic flow is expressed in passenger car units passing a point of the roa during a certain time interval. If not otherwise stated one hour is taken for this time interval and that usually the average evening peak hour.

The objective function is first used for the subproblem of the transport network optimization problem: defining the optimal dimension for a given traffic flow. Mathematically this is a simple problem. In this study the number of lanes is considered to be a continuous variable and the objective function is considered to be differentiable with respect to the number of lanes. So the differential calculus can be used to determine the optimum. Applying the differential calculus the necessary computations are simple when the shape of the objective function is simple. Such a simple shape is obtained when simple and similar shapes are used for the functions for the different elements of the objective function. These considerations have influenced the final definition of the functions for the elements of the objective function.

9.1.2 The Elements of the Objective Function

The objective function consists of the following elements:

(a)   evaluated travel time costs;
(b)   vehicle operating costs;
(c)   costs of accidents;
(d)   investments and maintenance of the road.

These costs are all expressed in the same monetary units. Furthermore all costs are discounted to present value. In this way it is also possible to compare more or less continuously appearing cost factors with more or less unique or periodically appearing cost factors.

A. Evaluated travel time costs

In defining this cost factor two factors play a role:

(a)   the defining of the total travel time as a function of the traffic flow and the dimension;
(b)   the social evaluation of the travel time.

The total travel time per year depends on the speeds at different traffic flows, the appearance of traffic-jams and so on. We try to approximate these travel times using as a base the relationship between speed and flow/capacity ratio or flow/dimension ratio (we will use the term flow/capacity ratio, because this is more common than the term flow/dimension ratio). We will here use a relationship in which the speed diminishes only at a rather high traffic density, but then rather steeply.

For every hour of the day and every day of the year the traffic flow is different and so also is the mean speed on the road. We assume now a particular distribution of traffic over the day and over the year—not basically different from the actual distribution to be observed at present—and use this to define the mean speed for every hour of the year and the contribution of the travel times of every hour to the total travel times per year. The total travel times per year are then computed, based on the traffic flow on the mean evening peak hour. It will be clear that the traffic flow in the evening peak hour is to a large extent decisive for the total travel times, of which evening peak hour has a larger share than any other hour. Furthermore in the evening peak hour the speed is at its lowest.

The evaluation of the travel time depends on the evaluation of the times spent at other activities and the (dis)pleasures of travelling.

B. Vehicle operating costs

The real vehicle operating costs are included in the objective function. They consist of fuel consumption and so on and that part of the depreciation which is directly caused by the use of the car, i.e. not the economic depreciation. The vehicle operating costs are assumed to depend on the flow/capacity ratio.

C. Costs of accidents

The material and evaluated non-material costs of accidents are included in the objective function. Again these costs are assumed to depend on the flow/capacity ratio.

D. Investments and maintenance

The construction costs for roads, intersections and other structures (bridges, tunnels) depend on the subsoil, the present building, the structure of the intersections and so on. Taking this into account for a number of types of road, intersection and structure, the relationship is defined between the construction costs and the number of (meeting) lanes. In these relationships the starting investments (up to 2 × 2 lanes) are generally rather high; expansion to 2 × 3 and 2 × 4 lanes is assumed to be comparatively cheap, while with further expansion 2 × 5 and 2 × 6 lanes again high investments are involved.

For all roads, intersections and other structures in the network, the type to which they (will) belong is defined. A number of lanes can already exist, for which there are of course no further construction costs. The costs for maintenance finally are assumed to depend on the number of lanes as well.

9.2 TRAVEL TIME COSTS

The treatment of the travel time costs consists of three elements:

(a) the social evaluation of the time spent travelling;
(b) the relationship between the mean speed of the traffic flow and the flow/capacity ratio;

(c)   the computation of the value of the total travel times per year based on a particular distribution of the traffic flows throughout the day and the year.

We will start this section by discussing travel time evaluation.

9.2.1   Travel Time Evaluation

We will here discuss travel time evaluation briefly. For a more extensive treatment see, for instance, Oort (1969) and de Donnea (1970).

According to the theory of consumer behaviour (see, for instance, Henderson and Quandt, 1958) every individual tries to maximize his own utility. In that case the marginal value of travel time equals the marginal value of time as an input for another activity (so, for instance, the income per unit time minus the evaluated displeasure of working) plus the evaluated displeasure of travelling for the time spent travelling.

From the above it follows that the value of travel time differs according to the individual and depends on the circumstances of the travelling. Because both the composition of the travellers and the circumstances of the travelling are related to the purpose of the travelling, i.e. the composition and the circumstances resemble more closely other compositions and circumstances with the same travel purpose than others with different travel purposes, the average value of travel time per purpose is often defined. Empirical research into the value of travel time losses tries to say something about this value on the basis of observed choices between different travelling possibilities with different values for travel (monetary) costs and travel times (see, for instance, Harrison and Quarmby, 1970). The general conclusion of this empirical research is, that the value of travel time differs per travel purpose and equals a certain part of the income per unit time.

De Donnea (1970) defines values for one hour travel time loss as shown in Table 9.2.1. Then he gives an estimate for the level and distribution of income

Table 9.2.1   Travel time evaluation according to De Donnea (1970) (Reproduced by permission of the Minister of Transport and 'Waterstaat', Nederlandse Staatsuitgeverij, the Netherlands)

Purpose	Value of one hour travel time loss
Home to work	30 to 40% of the spendable income per hour
Shopping and private business	30 to 40% of the spendable income per hour for the working population and the same amount for others
Social and recreational	20 to 30% of the spendable income per hour for the working population and the same amount for others
Business	The average wage costs per hour
Goods transport	The average wage costs of one hour's overtime for drivers times the average crew of a lorry

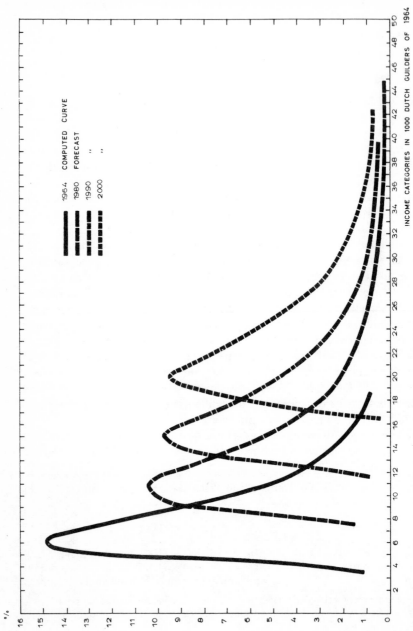

Figure 9.2.1 Income distribution of the working population older than 20 years for the Netherlands (forecast made by de Donnea, 1970. Reproduced by permission of the Minister of Transport and 'Waterstaat', Nederlandse Staatsuit-

for the years 1980, 1990 and 2000 (see Figure 9.2.1) and an estimate of the share of every income class in the total number of kilometres travelled per purpose. Taking these distributions into account he finds the amounts for the value of one hour travel time loss in car traffic as given in Table 9.2.2.

Table 9.2.2   Value of one hour travel time loss per person in car traffic (in Dutch guilders of 1970). (After de Donnea, 1970. Reproduced by permission of the Minister of Transport and 'Waterstaat', Nederlandse Staatsuitgeverij, the Netherlands.)

Purpose\Year	Home to work	Shopping and private business	Social and recreational	Business	Goods transport (per lorry)
1970	2·65	3·05	2·00	14·20	8·05
1980	3·40	4·15	2·60	21·50	12·60
1990	4·20	5·30	3·00	29·50	16·40
2000	5·00	6·60	3·55	41·50	20·95

Although according to the theory, only the value of the travel time loss or gain can be defined, this value is used for the whole travel time. So we write $k(z + \Delta z)$ for the travel time costs, while theoretically we must write $k'z + k\Delta z$, in which only k is known (z indicates travel time and k the value of travel time). The error made does not influence the results, however, because in the optimization only changes in the travel times are of relevance.

It has been pointed out already that the travel circumstances influence the evaluation of the travel time. The less comfortable these circumstances are, the more the travel time is felt. Because the comfort of driving a car depends for a large part on the flow/capacity ratio on a road (see for instance also the *Highway Capacity Manual 1965* (Highway Research Board, 1966)) a functional relationship between the value of travel time and the flow/capacity ratio is obvious. Unfortunately hardly any empirical research has been done on this subject, so it is difficult to define the correct relationship. For the sake of computational simplicity the following relationship is used:

$$k = \text{KZERO} + \text{KEXTRA}\left(\frac{x}{c}\right)^{\text{CCRE}} \qquad (9.2.1)$$

with

k—value of travel time

KZERO, KEXTRA, CCRE—coefficients

(In this second part of the book we will generally use capital letters or combinations of capital letters and figures to denote (scalar) coefficients.)

In this relationship the exponent CCRE equals the exponent in all other users' cost functions. The values for the coefficients KZERO and KEXTRA have been chosen in such a way that the resulting value of travel time agrees as far as possible with the values defined by de Donnea (see Table 9.2.2).

The values found by de Donnea are mostly based on data collected during the evening peak hours (see Section III and IV of Chapter I of de Donnea, 1970). So it seems correct to choose the values for the coefficient KZERO and KEXTRA such that at a traffic flow of 1250 p.c.u. per hour per lane (the average flow on the evening peak hour) the value of travel time equals the values of Table 9.2.2. Moreover, in the absence of sufficient empirical research it seems advisable to take KZERO not all that different from the values of Table 9.2.2. Taking this into account the following values for the coefficients KZERO and KEXTRA have been chosen:

$$\text{KZERO} = 0.9 \, k'$$

$$\text{KEXTRA} = 1.0 \, k'$$

with k'-value of travel time as given in Table 9.2.2.

Finally, one last remark must be made with respect to the character of the travel time cost and its evaluation. For the most frequently occurring travel purposes the value of travel time is defined on the basis of observed travellers' behaviour. The travellers respond to the costs as perceived and evaluated individually. It is not immediately true that this travel time evaluation equals the social travel time evaluation. However, in this study we state that the travel times, evaluated in the way described in this subsection, form a part of the social costs.

9.2.2 Relationship between Travel Times and Flow/Capacity Ratio

For the statement of the objective function it is necessary to know the travel time costs as a function of the (average) traffic flow and the dimension of the road. For this it is not important to know the exact speed realized by a particular car at a particular moment under particular circumstances, but an estimate must be made of the total travel times for a given average traffic flow and its distribution and for a given number of lanes.

To this end we use the relationship between the mean speed and the flow/capacity ratio. The *Highway Capacity Manual 1965* gives many of these relationships for all types of roads and intersections (see Figure 9.2.2). The relationships have been considered in the Netherlands by, among others, Beukers (1967) and Mulder (1971). In the highway capacity literature one usually speaks about the operating speed and the flow/capacity ratio of an ideal road under ideal conditions. The operating speed is defined as the highest overall speed at which a driver can travel under favourable weather conditions and under prevailing traffic conditions without at any time exceeding the safe speed. So the operating speed is a hypothetical speed which cannot be observed. For that very reason this concept seems to have been abandoned in favour of, for instance, the observable average speed. In the *Policy on Geometric Design of Rural Highways* (A.A.S.H.O., 1965b)

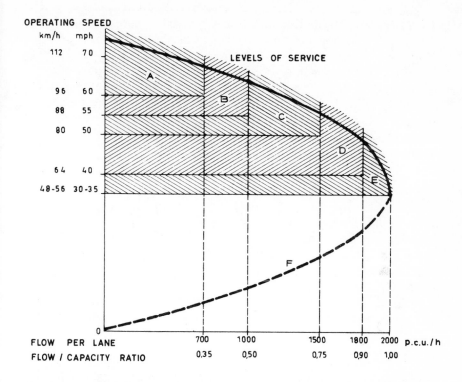

Figure 9.2.2 Relationship between operating speed and flow/capacity ratio on a 2 × 2-lane highway. Levels of service. (Reproduced from Beukers, *Ongelijkvloerse Kruispunten in Auto(snel)wegen* by permission of Nederlandse Wegencongres, 's-Gravenhage, the Netherlands.)

is stated that the average speed is about 8 to 10 km/hour lower than the operating speed.

An ideal road is defined as a road the lanes of which are not less than 3·60 m wide and which has adequate shoulders and no lateral obstructions within 1·80 m of the edge of the road-surface. Ideal conditions are defined as an uninterrupted flow, free from side interferences of vehicles and pedestrians, of passenger cars only and a good horizontal and vertical alignment with no restricted passing sight distances.

The operating speed is defined for a stable flow. So it must be considered as an upper bound for the speed. For an unstable flow the situation is different. When the speed falls, so the number of vehicles passing a point in the road per unit time decreases. The roles of dependent and independent variables switch and the question becomes what traffic flow is realized at a certain speed (the lowest dotted part of the curve in Figure 9.2.2).

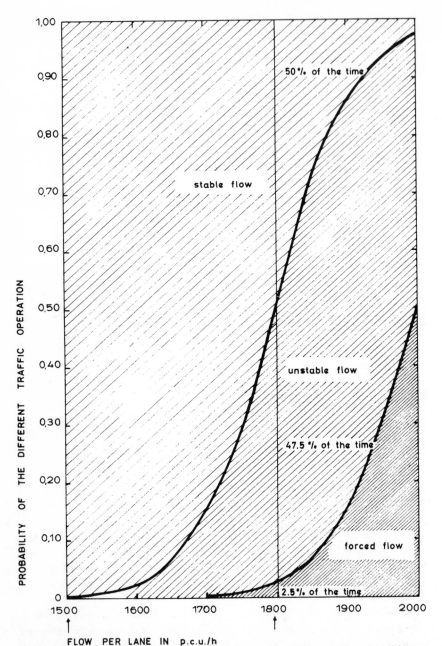

Figure 9.2.3 Relationship between traffic flow and traffic operation; conditions during the 5 minutes peak within the peak hour. (Reproduced from Beukers, *Ongelijkvloerse Kruispunten in Auto(snel)wegen* by permission of Nederlandse Wegencongres, 's-Gravenhage, the Netherlands.)

The probability of such instabilities in the traffic operation is again a function of the flow/capacity ratio (see Figure 9.2.3, taken from Beukers, 1967). In a traffic flow of 1,800 p.c.u. per hour per lane, the probability of an unstable flow is 50 per cent, while in a flow of 2,000 p.c.u. per hour per lane a forced traffic operation occurs in half the cases.

A decreasing flow for a decreasing speed causes a not-unique relationship between speed and flow/capacity ratio. For the optimization, however, unique relationships are necessary. Therefore we state a more or less artificial relationship between speed and supply of traffic flow.

How large the travel time losses are in situations with stagnations and unstable traffic operation depends on the total traffic flow over a longer period. To compute these travel time losses, techniques of queueing theory or simulation can be applied. Just to get an initial idea of the amount of the travel time losses we will give a simplified computation without applying the advanced techniques of queueing theory or simulation. Suppose at a total supply of 2,500 p.c.u. per hour per lane stagnation occurs. Suppose further, that the first 1,000 p.c.u. can continue with a speed of 14 km per hour. So the next 1,000 p.c.u. must wait about 4 minutes before they can continue with a speed of 14 km per hour. The mean speed of these 1,000 p.c.u. is then about 7 km per hour. The last 500 p.c.u. must wait about 8 minutes and the mean speed of these is about 5 km per hour.

Observations of the mean speed at different flow/capacity ratios, on the basis of which the shape of the speed/flow relationships and its coefficients could be estimated and tested with the help of statistical techniques, are almost absent, at least for the area with higher flow/capacity ratios. There-fore some points of the curve have been chosen arbitrarily in consultation with traffic engineers at the Rijkswaterstaat and the Netherlands Economic Institute. Next a plausible form for the relationship has been chosen and the coefficients have been defined such that the best agreement with the chosen points has been obtained. The following functional relationship has been applied:

$$z = \text{TZERO} + \text{CCR}\left(\frac{x}{c}\right)^{\text{CCRE}} \qquad (9.2.2)$$

Different values for the exponent CCRE have been chosen and for all these values the 'best' values for TZERO and CCR are defined (using the method of least squares). There is almost no difference in the squared dis-crepancies for the values 5, 6 and 7 for the exponent CCRE (the correlation coefficient (R^2) is respectively 0.985, 0.995 and 0.998). Traffic engineers of the Rijkswaterstaat and the Netherlands Economic Institute think that the value 7 for the exponent CCRE gives the best description of reality. This high value for the exponent CCRE gives a flatter course than the relationship given

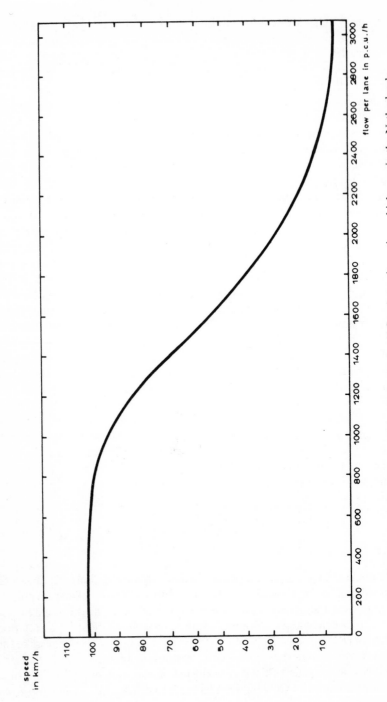

Figure 9.2.4   Relationship used between speed and flow/capacity ratio on highways in the Netherlands

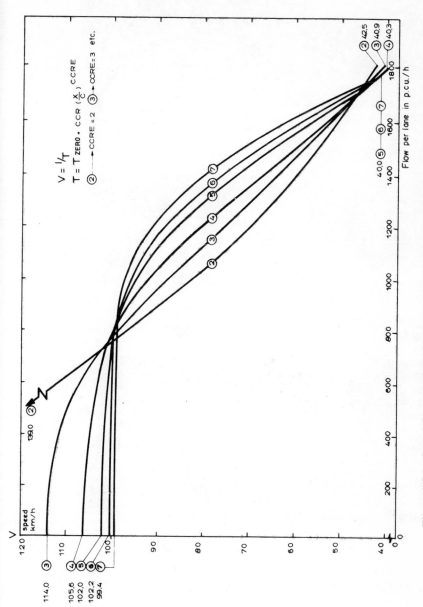

Figure 9.2.5 Speed/flow relationships determined on the basis of observations on Dutch highways

A
0 - 700
p.c.u. ph. p.l.

B
700 - 1000
p.c.u. ph. p.l.

C
1000 - 1500
p.c.u. ph. p.l.

Figure 9.2.6   Levels of service. (Reproduced from *Highway Capacity Manual, 1965* (pp. 84, 85) by permission of Highway Research Board, National Academy of Sciences, Washington D.C.)

D
1500-1800
p.c.u. p.h. p.l.

E
1800 - 2000
p.c.u. p.h. p.l.

F
< 2000
p.c.u. p.h. p.l.

in the *Highway Capacity Manual* for the area with low flow/capacity ratios. Such a flat course is also urged by Schlums (1967).

On the other hand there are computational advantages, when for all different users' cost factors, such as travel time costs, vehicle operating costs and costs of accidents, similar functional relationships between costs and flow/capacity ratio are used. Using the form $x\{\alpha + \beta(x/c)^{CCRE}\}$ for all cost factors, a value for CCRE lower than 7 seems preferable. For this reason the value 5 has ultimately been chosen for the exponent CCRE, in view also of the small differences between the different curves for the values 5, 6 and 7 for CCRE (see Figure 9.2.5) and of the uncertainty with respect to the correct functional relationship. So the following relationship between the travel time in hours per kilometre and the flow/capacity ratio has been used:

$$z = \text{TZERO} + \text{CCR}\left(\frac{x}{c}\right)^{CCRE}$$

with

$$\text{TZERO} = 0.00980$$

$$\text{CCR} = 0.02564$$

$$\text{CCRE} = 5$$

The function is shown in Figure 9.2.4.

So we use the same relationship between travel time and flow/capacity ratio for all different types of roads (2-lane and more-lane). Though this does not accord with reality, it is obvious that the use of more travel time relationships would have made the computations more complicated (see also Section 10.1.3).

It is interesting to go somewhat deeper into the effect of the value for the exponent CCRE. It is obvious that when a higher value for CCRE is used, the speed falls only at a higher value for the flow/capacity ratio, but on the other hand the speed is falling more steeply and finishes lower (two intersections of the different travel time curves). An irregular distribution of the traffic with many high peaks and low valleys, has a much greater impact at a high value of CCRE. The influence of the exponent CCRE on the total travel time costs per year, and thus on the optimal relationship between traffic flow and number of lanes under different assumptions for the distribution of the traffic during the day and the year, will be treated in greater detail in Section 10.2.

The speed on the road is a measure of the quality of the traffic operation. The quality depends on the average speed but also on factors such as the total travel time, traffic interruptions and restrictions, the freedom to manoeuvre, safety, driving comfort and convenience, whether there are traffic jams or not and so on. To indicate the quality of the traffic operation, in traffic engineering, so-called 'levels of service' are used (see the *Highway*

Capacity Manual 1965). These levels of service depend on the road type, the flow/capacity ratio and the actual speed, if the operating speed is not assumed to be the only possible speed. These relationships are shown in Figure 9.2.2.

Six levels of service are distinguished, with the following characteristics:

1. Level of service A (0–700 p.c.u. per hour per lane for multilane highways under ideal conditions): condition of free flow with no restriction on manoeuvrability and full freedom of speed.
2. Level of service B (700–1,000 p.c.u. per hour per lane): stable flow with operating speeds beginning to be restricted somewhat by traffic conditions; still reasonable freedom for the drivers to select their speed and lane of operation.
3. Level of service C (1,000–1,500 p.c.u. per hour per lane): stable flow but speeds and manoeuvrability are more closely controlled by traffic flows; restricted freedom to select own speed, to change lanes or to overtake.
4. Level of service D (1,500–1,800 p.c.u. per hour per lane): unstable flow is approached; operating speeds are considerably affected by changes in operating conditions; fluctuations in volume and temporary restrictions to flow may cause substantial drops in speeds; little freedom to manoeuvre.
5. Level of service E (1,800–2,000 p.c.u. per hour per lane): unstable flow, low speed and possible stoppages of momentary duration.
6. Level of service F (Less than 2,000 p.c.u. per hour per lane): forced flow, very low speed, forming of queues and stoppages for short or long periods of time.

The visual picture of the traffic flows represented by these levels of service is given in Figure 9.2.6.

In the computations the relationship between speed and flow/capacity ratio is assumed not to change during the period 1970–2000. This has been based on the fact that speed is not determined in the first place by the technical possibilities of the vehicle or the road, but by the abilities of and constraints upon the driver (van der Burght *et al.*, 1970). Only by automation of the traffic operation could much higher speeds and densities be possible. However, before these developments can be taken into account in defining the optimal road network, further research is needed into the technical and the economic possibilities.

9.2.3 Travel Time Costs for the Whole Year

The relationship between travel time and flow/capacity ratio discussed in Section 9.2.2 in combination with the travel time evaluation discussed in Section 9.2.1 is used to compute the total travel time costs per year based on the average flow.

The flow/capacity ratio varies over the day and over the year. This means that every hour makes a different contribution to the total travel time costs per year. Moreover the speed is different per hour and so is the value of travel time due to changes in comfort at changing flow/capacity ratio. Finally the composition of the traffic is not constant over the day and over the year. In summer there is, for instance, more recreational traffic and, during the peak hour, commuter traffic is important. This implies that the average value of travel time also varies per hour. When the traffic flow and average value of travel time are known per hour, it is possible to compute the total value of travel time costs per year as follows:

$$F^{\text{TIME}} = \sum_{h \,\in\, \text{all hours of the year}} x_h z_h k_h \tag{9.2.3}$$

Substituting the relationships (9.2.1) and (9.2.2) into equation (9.2.3) we get:

$$F^{\text{TIME}} = \sum_h x_h \left\{ \text{TZERO} + \text{CCR}\left(\frac{x_h}{c}\right)^{\text{CCRE}} \right\}$$

$$\times \left\{ \text{KZERO}_h + \text{KEXTRA}_h\left(\frac{x_h}{c}\right)^{\text{CCRE}} \right\}$$

$$= \sum_h x_h \text{TZERO} \cdot \text{KZERO}_h$$

$$+ \sum_h (\text{CCR} \cdot \text{KZERO}_h + \text{TZERO} \cdot \text{KEXTRA}_h) x_h \left(\frac{x_h}{c}\right)^{\text{CCRE}}$$

$$+ \sum_h x_h \text{CCR} \cdot \text{KEXTRA}_h \left(\frac{x_h}{c}\right)^{2\text{CCRE}}$$

We express the total travel time costs per year as a function of the flow on a base hour:

$$F^{\text{TIME}} = x_b \sum_h \text{TZERO} \cdot \text{KZERO}_h \frac{x_h}{x_b}$$

$$+ x_b \left(\frac{x_b}{c}\right)^{\text{CCRE}} \sum_h (\text{CCR} \cdot \text{KZERO}$$

$$+ \text{TZERO} \cdot \text{KEXTRA}_h)\left(\frac{x_h}{x_b}\right)^{\text{CCRE}+1}$$

$$+ x_b \left(\frac{x_b}{c}\right)^{2\text{CCRE}} \sum_h \text{CCR} \cdot \text{KEXTRA}_h \left(\frac{x_h}{x_b}\right)^{2\text{CCRE}+1}$$

or, with new coefficients:

$$F^{\text{TIME}} = x_b \left\{ \text{D2TIME} + \text{D1TIME} \left(\frac{x_b}{c} \right)^{\text{CCRE}} + \text{D3TIME} \left(\frac{x_b}{c} \right)^{2\text{CCRE}} \right\} \quad (9.2.4)$$

with

F^{TIME} —total travel time costs per year per unit of length

x_h —traffic flow for hour h in p.c.u.

x_b —traffic flow for the base hour in p.c.u.

k_h —average value of travel time per p.c.u. for hour h

KZERO_h —$0 \cdot 9\, k_h'$ (see Section 9.2.1)

KEXTRA_h—$1 \cdot 0\, k_h'$ (see Section 9.2.1.)

D1TIME, D2TIME, D3TIME—coefficients to be computed.

For the flow for the base hour we can take the average annual hourly flow, the flow for the average peak hour or for the decisive hour. Because the exponent CCRE is generally larger than one (five in our case), the distribution of the traffic influences the values of the constants D1TIME and D3TIME to a large extent. Based on the average annual hourly flow, a flat distribution of the traffic over time gives a lower value for D1TIME and D3TIME than a distribution with many peaks and valleys. Even without the formulae above it is obvious that with the same dimension for the road the total time costs are less with a flat distribution of traffic (consider the benefits of spreading the peak hour out). Taking the average or decisive peak hour as base hour a flat distribution gives a high value for the constants DITIME and D3TIME. However, this is just a matter of definition, for with a flat distribution many hours will have the same or nearly the same traffic flow as the peak hour. The effect of the different distributions over time of the traffic will be further discussed in Section 10.2.

Taking the peak hour as base hour, the quotient x_h/x_b will generally be smaller than one; this implies that D3TIME will be small with respect to D1TIME. Because the quotient x_b/c will also generally be smaller than one, the last term in relationship (9.2.4) for the travel time costs per year will be much smaller than the first ones. So it seems justified to neglect this last term.

The quotient D1TIME/CCR can be interpreted as the total value of travel time loss per year when on the base hour one p.c.u. experiences a travel time loss of one hour. For, if the ratio x_b/c is such that on the base hour Δz hours travel time loss occurs, the total value per year of all travel time losses ΔF^{TIME} can be computed as follows:

$$\Delta z = \text{CCR} \Delta \left(\frac{x_b}{c} \right)^{\text{CCRE}} \quad (9.2.5)$$

$$\Delta F^{\text{TIME}} = x_b \text{D1TIME} \Delta \left(\frac{x_b}{c} \right)^{\text{CCRE}} \quad (9.2.6)$$

or substituting equation (9.2.5) into equation (9.2.6):

$$\Delta F^{\text{TIME}} = x_b \frac{\text{D1TIME}}{\text{CCR}} \Delta z \qquad (9.2.7)$$

The quotient D2TIME/TZERO gives the value per year of a difference in travel time for the base hour if the speed and the value of travel time were independent of the flow/capacity ratio.

In defining the distribution of the traffic over the day, over the week (work days versus Saturday and Sunday) and over the year and the distribution over the different travel purposes at each moment, many factors are important. The most important factor, however, is the length of the time devoted to work. For the Dutch Integral Transportation Study this subject was studied by Hendriks and Vloemans (1971). They point out that there are many ways of shortening time devoted to work: shortening the working day or working week, lengthening holidays and shortening a man's working life (longer schooling and/or earlier retirement). Each of these possibilities would have its own impact on the distribution of traffic and thus on the optimal infrastructure. Based on expectations for the future and on known present distributions (observations of the Rijkswaterstaat and results of the home interview survey in 1966 by the Committee for the Promotion of Public Transport in the West of the Netherlands (C.O.V.W., 1966)), Hendriks and Vloemans give distributions over time for the traffic by car for different travel purposes.

The distribution over the different purposes and between goods and personal transport is defined in the transport forecast. The ultimate results of the Dutch Integral Transportation Study not being available at that time, the distribution has been based on figures given by Heijke (1970), estimated on the basis of the C.O.V.W.-survey and figures of the Centraal Bureau voor de Statistiek.

Combination of these two distributions yields a complete distribution of the traffic over the year and of the average value of the travel time (x_h and k_h for every hour of the year). Some important results are shown in Figure 9.2.7 and 9.2.8.

It has been said already, earlier in this subsection, that the distribution of the traffic has a great impact on the travel time costs and thus on the necessary infrastructure. Therefore it is important to have a good insight into the developments influencing this distribution. Moreover it is possible to try to use the distribution over time of the traffic as an instrument to regulate transport. For these reasons and because of the considerable uncertainty about future developments of the distribution over time, the distribution for 1980 has also been used for the years 1990 and 2000 by way of precaution. The distribution of 1980 also most resembles the present one, in which,

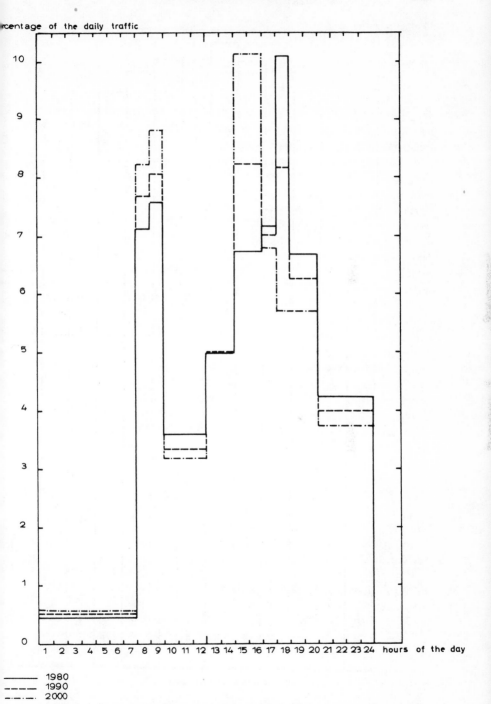

Figure 9.2.7 Supposed distribution of person traffic by car over the hours of the
average work day (work day in October)

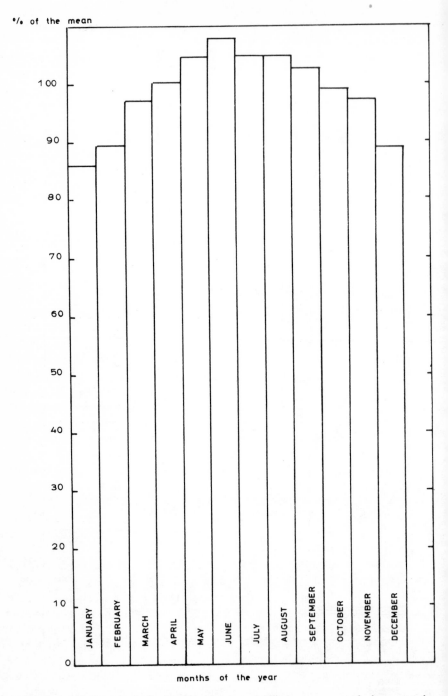

Figure 9.2.8   Supposed distribution of person traffic by car on work days over the months of the year (1980)

according to the Rijkswaterstaat, up to now no systematic change has occurred.

Another important factor is the number of occupants per car. For this factor again estimates by Heijke (1970) have been used. The value of travel time per person from Table 9.2.2 is in this way recomputed to a value of travel time per p.c.u. These values are given in Table 9.2.3.

Table 9.2.3   Value per car of one hour travel time loss (in Dutch guilders of 1970)

Year	Purpose Home– work	Business	Remaining	Goods transport	On Saturdays	On Sundays
1980	4·08	23·65	5·20	12·60	6·16	5·99
1990	4·62	32·45	6·39	16·40	7·33	6·70
2000	5·50	45·65	6·58	20·95	7·31	7·67

The values for k_h and x_h being known, it is possible to compute the coefficients D1TIME and D2TIME for the different years. Having the values for these coefficients, the total travel time costs per year can be computed based on the flow for the base hour. Taking the same value for D1TIME and D2TIME for all roads, it must be assumed that the distribution of the traffic on all roads is the same. This is certainly not always true. In particular large regional and local differences exist in the split between the traffic on work days and on Saturday and Sunday. These differences are so large that it does not seem permissible to use the forecasted traffic flow on *one* base hour to compute the travel time costs for the whole year (see Section 6.1.4 for the theoretically correct tackling of this problem).

The distributions over time for work day traffic vary less geographically. So it seems permissible to compute the total travel time costs per year on work days based on a forecast of the flows for one base hour with a small correction discussed later on in this subsection. In conclusion, we see that the objective function contains the total travel time costs (and other users' costs) per year on workdays.

Taking into account the variations over time per hour of the day and per month of the year (Figure 9.2.7 and Figure 9.2.8) we neglected variations within the hour, over the different (work)days of the week and over the different weeks of the month. To make an estimate for the variations within the hour, traffic flows on five-minute intervals within the peak hour given in the *Highway Capacity Manual 1965* have been used. Also for Dutch highways the peak within the peak hour has been determined (see Table 9.2.4, taken from Beukers, 1967). Ultimately the distribution of Figure 9.2.9 has been used as an estimate for the distribution of the traffic within every hour.

Table 9.2.4   Peak factor on Dutch highways in 1967 (Reproduced by permission of Nederlandse Wegencongres, 's-Gravehage from Beukers, B. Vormgeving van ongelijk-vloerse kruispunten, Preadvies Congresdag 1967: *Ongelijkvloerse Kruispunten in Auto(snel)wegen*).

Rijksweg (highway) nr.	Location	Traffic flow for the five-minute peak as a percentage of the total flow for the evening peak hour
2	Near Breukelen	12·8
4	Near Hoofddorp	11·2
12	Near Oudenrijn	11·2
13	Between Delft and Overschie	9·7
16	At the Van Brienenoordbrug	9·7

For the fluctuations within the week the distribution of Figure 9.2.10 has been used, also extracted from data of the *Highway Capacity Manual 1965*. Again, for the weeks within the month, variations in the traffic flow occur. To make an estimate for these, it has been assumed that the difference in flows between two succeeding months is built up smoothly during the succeeding weeks of the months. This leads to the distribution over the weeks for the average month given in Figure 9.2.11.

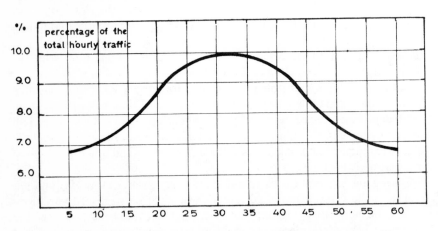

Figure 9.2.9   Distribution of traffic within the hour

As said before the distribution over time of the traffic flow is different for each road. See Figure 9.2.12 for four rather different distributions over the day on four Dutch highways. Still the same distribution is assumed on all

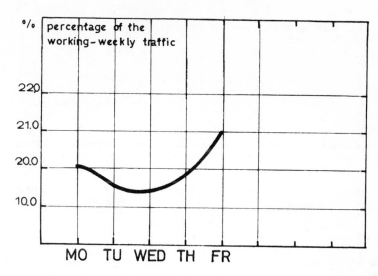

Figure 9.2.10   Distribution of traffic over the days of the week

roads. So it is necessary to use a correction factor to neutralize the error made. This correction factor is based on a distribution defined for the geographical variations in the time distributions based on data from the Rijkswaterstaat and the *Highway Capacity Manual 1965*. The distribution is shown in Figure 9.2.13.

All these distributions influence the travel time and the value of travel time. Because of the non-linear relationships between these two factors and

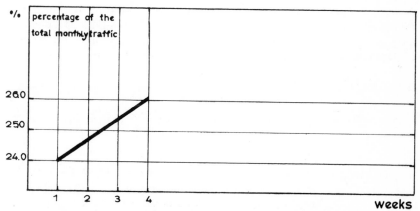

Figure 9.2.11   Distribution of traffic over the weeks of the month

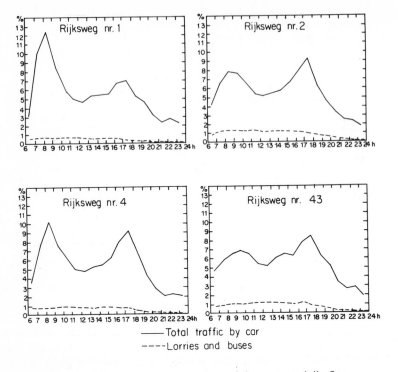

————Total traffic by car

————Lorries and buses

Figure 9.2.12 Hourly flows in percentage of the average daily flow per direction for car traffic in 1970 on some roads in the Netherlands. (Reproduced by permission of the Minister of Transport and 'Waterstaat', the Netherlands, from Rijkswaterstaat, Dienst Verkeerskunde. *Verkeerstellingen in 1970*).

the flow/capacity ratio the mean travel time costs are higher than the travel time costs of the mean traffic flow. This implies an adjustment of the factor D1TIME as computed up to now. Note that the factor D2TIME does not need to be adjusted, because this one is only linearly dependent on the flow. For D1TIME we can write down the following relationship:

$$D1TIME = D1TIME'$$

$$\times \int_0^\infty \{F_1(x)F_2(x)F_3(x)F_4(x)\} \, d(x^{CCRE+1}) \qquad (9.2.8)$$

with:

D1TIME'—D1TIME without taking into account the variations within the hour, the week and the month and the geographical differences in the distributions over time

$F_i(x)$ —distribution of x within i (with $\int_0^\infty F_i(x) \, dx = 1$)

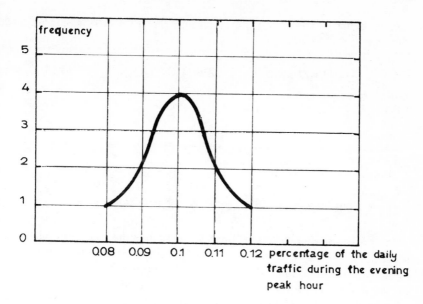

Figure 9.2.13 Geographical differences in the distributions of traffic

For the distributions given in the Figures 9.2.9, 9.2.10, 9.2.11 and 9.2.13 the value of the integral in relationship (9.2.8) is 1.6771. It may be of interest to show the contribution of the different variations. To this end we look at the factor

$$\left\{ \int_0^\infty F_i(x) \, dx^{\text{CCRE}+1} \right\}^{1/(\text{CCRE}+1)}$$

which can be interpreted as a factor by which the average flow must be multiplied to take into account the related variations. For the variations within the hour, the week, and the month and the geographical differences, this factor is respectively 1·050, 1·004, 1·003 and 1·031.

The discussions in this section lead to values for the coefficients D1TIME and D2TIME in the relationship (9.2.9), which are used to compute the total travel time costs per year (for workdays) based on the traffic flow on the average evening peak hour:

$$F^{\text{TIME}} = x_b \left\{ \text{D2TIME} + \text{D1TIME}\left(\frac{x_b}{c}\right)^5 \right\} \tag{9.2.9}$$

The values vary per year due to changes in real income and the number of occupants of a car (working *via* the travel time evaluation). These values are given in Table 9.2.5, in which for reasons of completeness D1TIME has been

Table 9.2.5   Values for the coefficients for the travel time costs relationship (in Dutch guilders of 1970 per kilometre)

Year	$\sum_h CCR.KZERO_h\left(\dfrac{x_h}{x_b}\right)^6$	$\sum_h TZERO.KEXTRA_h\left(\dfrac{x_h}{x_b}\right)^6$	D1TIME	D2TIME
1980	94·97	40·56	135·53	124·71
1990	111·11	47·44	158·55	145·91
2000	123·05	52·56	175·61	161·59

split into

$$\sum_h CCR.KZERO_h\left(\frac{x_h}{x_b}\right)^6 \quad \text{and} \quad \sum_h TZERO.KEXTRA_h\left(\frac{x_h}{x_b}\right)^6.$$

In this section there has been ample discussion of the relationship between speed and flow/capacity ratio in order to determine the total travel time costs per year on a road, based on the average traffic flow on that road and its distribution. However, there are other ways of determining the total travel time costs per year. One interesting method has been used in Nordrhein-Westfalen in Germany (Behrendt and Kloss, 1970). In this method the costs are determined on the basis of the probability of congestion, leading to a certain travel time loss, as a function of the flow/capacity ratio. Working out this idea gives the following relationships for the total travel time costs per year:

$$F^{\text{TIME}} = \sum_h \sum_i p_i\left(\frac{x_h}{c}\right)\tau_i(x_h) \tag{9.2.10}$$

with

$p_i(x_h/c)$—probability of congestion of type i as a function of the flow/capacity ratio

$\tau_i(x_h)$ —total travel time costs of a situation with congestion of type i as a function of the flow

Of course it is possible to use combinations of these and other approximations to determine the total travel time costs.

9.3   VEHICLE-OPERATING COSTS

The costs connected with the possession and use of a car can be divided into two categories:

(a)   ownership costs;
(b)   running costs.

As with all other cost factors, the vehicle operating costs must be taken into the objective function without taxes. For only the real social costs are involved and transfers in the form of taxes are not relevant. We emphasize this here because the share of the taxes in the market-prices of these cost factors are very high (in the Netherlands, for instance, about 75 per cent for petrol).

9.3.1 Ownership Costs

The ownership costs are composed of the following cost factors:

(a) purchasing costs or loss of interest;
(b) insurance contributions (these costs are partly considered as costs for accidents);
(c) costs for garage accommodation;
(d) costs for maintenance;
(e) depreciation as far as economic ageing is concerned;
(f) other cost factors.

These cost factors are all virtually independent of the number of kilometres travelled and are not therefore relevant to defining the optimal road network. They must be included in the objective function as constants, but can then also be neglected.

9.3.2 Running Costs

The running costs consist of the following elements:

(a) use of fuel;
(b) grease and use of oil;
(c) wear on tyres;
(d) other repairs;
(e) depreciation as far as technical ageing is concerned.

The level of the different cost factors depends rather on the type of car (the engine capacity is decisive) and other factors. So an 'average car' and an 'average user' must be assumed. We take a car with an engine capacity of about 1,250 cc, driving about 20,000 km per year. To estimate the level of the different cost factors data have been used per type of car provided by the Royal Dutch Tourist Association A.N.W.B. (1970). These cost factors are given in Table 9.3.1.

For this definition of the costs, half of the costs for repair (excluding wear on tyres) and two thirds of the total depreciations are assumed to be independent of the number of kilometers travelled.

Furthermore these costs are assumed not to change over time. Without essential changes in the car one can expect, on the one hand a decrease of the costs owing to a more efficient use of fuel etc., and on the other an increase of

Table 9.3.1   Running costs without taxes (in Dutch guilders of 1970)

Cost factor	Cost per kilometre
Use of fuel	0·013
Grease and use of oil	0·005
Wear of tyres	0·011
Other repairs (50% of total)	0·010
Depreciation ($33\frac{1}{3}$% of total)	0·011
Total	0·050

costs owing to the use of more expensive fuel (perhaps also to measures taken to diminish the costs to the environment) and to higher wage rates.

For the running costs there also exists a certain relationship between cost and flow/capacity ratio. In high traffic densities and when queues are formed, braking and accelerating are more often called for. This causes extra use of fuel, extra wear on tyres and extra wear on the gearbox, clutch and so on. The total extra costs for driving in congested traffic conditions are estimated at about 25 per cent by the A.N.W.B. Almost the same percental increase of the running costs after an increase of the flow/capacity-ratio is mentioned by Soberman and Clark (1970).

For computational reasons the same functional relationship has been chosen for the running costs as for the travel time costs:

$$\text{D2RUN}' + \text{D1RUN}'\left(\frac{x}{c}\right)^5 \tag{9.3.1}$$

The coefficients D2RUN′ and D1RUN′ have been chosen in such a way that the average running costs equal the 5 Dutch cents of Table 9.3.1 and that these costs are at a flow/capacity ratio of 1,500 p.c.u. per hour per lane about 25 per cent higher than for an unloaded road. Per driven kilometre D1RUN′ equals then 4·81. Dutch cents and D2RUN′ 4·5 Dutch cents.

To compute the total running costs per year based on the traffic flow for the average peak hour, the following relationship is used:

$$F^{\text{RUN}} = x_b\left\{\text{D2RUN} + \text{D1RUN}\left(\frac{x_b}{c}\right)^5\right\} \tag{9.3.2}$$

Table 9.3.2   Values for the coefficients for the running costs relationship (in Dutch guilders of 1970 per kilometre)

Year	D1RUN	D2RUN
1980/1990/2000	34·60	112·83

o determine the values for the coefficients D1RUN and D2RUN in this
rmula, based on the costs per kilometre, the same computations must be
ecuted, as discussed in Section 9.2.3. Of course the same distributions
ver time for the traffic are used as in Section 9.2.3. These computations yield
e values given in Table 9.3.2 for the coefficients D1RUN and D2RUN
determine the total running costs per year (for traffic on workdays).

9.4 COSTS OF ACCIDENTS

Defining the costs of accidents consists of three parts:

) the definition of the social costs per type of accident;
) the definition of the number of accidents of a certain type per driven
kilometre;
the definition of the possible relationship between the number of
accidents of a certain type and the flow/capacity ratio.

efore going into the definition of the costs of accidents, we should point
t two other cost factors in which a part of the costs, related to the risks and
ngers of car-travelling has been included. These are firstly the extra costs
highway construction and traffic management installations etc. increasing
fety on the road and secondly the decrease of comfort and security ensuing
on the risk of accidents which is expressed in the definition of the value of
vel time.
For the definition of the social costs of accidents for the Dutch Integral
ansportation Study, a study of de Buin's (1970) has been used, which is
sed again on a number of other publications (among others, Winch, 1963;
ederlands Vervoerswetenschappelijk Instituut, 1950; Stichting Weten-
happelijk Onderzoek Verkeersveiligheid, 1963). De Bruin gives the survey
Table 9.4.1 for the costs of different types of accidents. For the definition
the economic costs of a person being killed there are several possibilities
e also the discussion of the Ninth Round Table, Paris 1970, Beesley and
ans, 1970). A method often used is the so-called net method (Nederlands
rvoerswetenschappelijk Instituut 1950; Stichting Wetenschappelijk On-
rzoek Verkeersveiligheid, 1963; Beesley and Evans, 1970).
According to this method the total costs equal the present value of the
fference of the goods not produced and not consumed by the victim.
is method raises the paradox that the killing of an aged man or that of a
ung child when the discount rate is sufficiently high means an economic
nefit. In the net method the society is assumed to consist of the next of kin.
e gross method also applied by de Bruin seems better. In this method the
tims are still considered as members of society so that the drop in con-
mption does not imply an economic profit. The present value of the goods
t produced are considered as an economic loss.

Table 9.4.1   Costs of accidents in 1968 according to de Bruin (1971) (in Dutch guilde of 1968) (Reproduced by permission of the Minister of Transport and 'Waterstaa Nederlandse Staatsuitgeverij, the Netherlands)

Cost factors	Cost per person killed	Costs per person wounded	Cost of material damage onl per accident
Economic cost per person killed/wounded	200,000	3,000	
Subjective cost per person killed/wounded	25,000	1,150	
Damage to vehicle	$1,800^a$	300^a	300^a
Damage to fixed objects	50	50	50
Damage to clothes, goods etc.	150	150	70
Costs of police	65	65	65^c
Costs of the court	45	45	45^c
Costs of handling	160	160	80
Extra congestion	600^b	400^b	100^b

[a] Times the number of concerned vehicles
[b] Own estimate
[c] If registered

Table 9.4.2   Number of accidents and costs of accidents per driven kilometre (Aft de Bruin, 1971. Reproduced by permission of the Minister of Transport and 'Wate staat', Nederlandse Staatsuitgeverij, the Netherlands)

Year	Road-type	Number of accidents	With a fatality	With serious injury	With light injury	Only with material damage	Costs pe driven kilometr in Dutcl guilders 1968[b]
		Per 10^6 driven kilometres					
1966	1 × 2	3·61	0·08	0·66	0·24	2·63	0·026
1966	2 × 2	1·71	0·02	0·18	0·10	1·40	0·009
1968[a]	1 × 2	1·4	0·11	0·37			0·29
1968	2 × 2	1·1	0·02	0·3			0·10
1968	2 × 3	0·7	0·0	0·18			0·003
1968 (on Sun-days)	2 × 3	1·4	0·01	0·65			0·006

[a] On 1 January 1967 the official reporting and so the registration changes: no official repo is made for material damage below 1000 Dutch guilders per vehicle.
[b] In defining the costs, corrections are used for non-registered accidents.

Next de Bruin gives an estimate for the development of the different ɔsts of accidents over time, related to the increase in real income.

For the number of accidents per driven kilometre in the Netherlands e Bruin gives figures, based on data of the Stichting Wetenschappelijk ɔnderzoek Verkeersveiligheid and the Rijkswaterstaat (see Table 9.4.2).

Finally, the relationship between the number of accidents per driven ilometre (accident ratio) and the flow/capacity ratio is investigated. De Bruin ɔplies regression analysis to accident data of the Stichting Wetenschappelijk ɔnderzoek Verkeersveiligheid and the Rijkswaterstaat for the Netherlands. e observed a slight correlation between accident ratio and flow/capacity tio but no correlation between the ratio of serious accidents and flow/ ιpacity ratio. This last ratio is of course very important for accident costs.

This absence of any relationship between accident ratio and flow/capacity tio is confirmed by other research (for instance Adam, 1961 and Winfrey, ι69). However, it is denied by still other research (for instance Pfund and ɔoenmerer, 1970; Behrendt and Kloss, 1970; Gwynn, 1967). These last ιdies observe a u-shaped relationship. Gwynn analyses the accidents on .S. Route 22 on the basis of accident data per hour for the years 1959 to ι63. At a flow of 1,500 p.c.u. per hour per lane he observes two to three times many accidents per driven kilometre as at 750 p.c.u. per hour per lane. e also observes a u-shaped relationship between the number of persons ɔunded per driven kilometre and the flow/capacity ratio; but this relation- .ip is less sharp. Recent research by the traffic research department of the ιjkswaterstaat (1972) leads to a similar relationship between accident tio and level of service on highways in the Netherlands in 1971 (see Table 4.3). A positive correlation between the ratio of serious accidents and the vel of service was also observed.

All in all it has proved difficult to derive a unique relationship between cident ratio and flow/capacity ratio. Yet it seems necessary to include the sts of accidents in the objective function and to let them influence the finition of the optimal dimension, in view of the great importance the jkswaterstaat attaches to the costs of accidents at the point of defining e permissible flow/capacity ratio. To that end the following relationship

Table 9.4.3 Accident ratios on highways in the Netherlands at different levels of service (in 1971)

Level of service	Number of p.c.u. per hour per lane	Number of accidents per driven kilometre (index)
A or B	0–1,000	100
C	1,000–1,500	150
D	1,500–1,800	162
E or F	>1,800 or congestion	200

for the accident ratio is used:

$$D2ACC' + D1ACC'\left(\frac{x}{c}\right)^5 \qquad (9.4.$$

For D2ACC' 0·5 Dutch cents per kilometre has been taken and for D1ACC
2·1 Dutch cents.

To compute the total costs of accidents per year (on workdays) the follow
ing relationship is used:

$$F^{ACC} = x_b \left\{ D2ACC + D1ACC\left(\frac{x_b}{c}\right)^5 \right\} \qquad (9.4.2$$

The coefficients D1ACC and D2ACC have been determined in the usu.
way, taking into account the distribution of the traffic over time (compar
Section 9.2.3 and 9.3.2). Furthermore the values of these coefficients diffe
per year due to the increase in real income. The values used for the coefficien
are given in Table 9.4.4.

Table 9.4.4   Values for the coefficients for the accidental costs
relationship (in Dutch guilders of
1970 per kilometre)

Year	D1ACC	D2ACC
1980	19·80	16·14
1990	24·75	20·18
2000	29·70	24·22

In the above it has become clear that it is very difficult to define th
dependence of accident costs on the traffic flow and the number of lanes of th
road. For computational considerations the relationship discussed, betwee
costs of accidents and flow/capacity ratio in the average evening peak hou
has been chosen. The coefficients in this relationship have been determin
as well as possible. Comparison of the values for these coefficients with tho
for the coefficients for the travel time and running costs relationships sho
that the influence of the factor D1ACC on the optimal flow/capacity rat
is rather small. Therefore it may be good to point out here that safety on th
road is a very important factor for the Rijkswatersaat when defining th
permissible flow/capacity ratio. The principle used is that stoppage of t
traffic flow on highways must be avoided at all costs, because stoppage do
greatly increase the likelihood of accidents occurring. From Figure 9.2.3
follows that the likelihood of stoppage begins to rise at a flow of 1,500 p.c
per hour per lane.

9.5 INVESTMENTS AND MAINTENANCE

9.5.1 General Remarks

The cost factor, investments and maintenance for road infrastructure, is divided into four parts:

(a) investments in roadway sections;
(b) investments in intersections;
(c) investments in special structures;
(d) costs for maintenance.

The cost price of roadway sections is composed of the following elements:

(a) land acquisition and adaptation of existing construction;
(b) earthworks;
(c) paving;
(d) 'normal' structures in and over the road;
(e) other elements.

This division may be a start for the composition of a cost model, in which the cost price of each element will be a function of the ultimate dimension and such variables as alignment, existing construction, basis and so on. In order to define these costs many optimization problems must be solved again, for the minimal costs must be given. This does not imply underestimation of the different cost factors, but such a choice of the alignment and so on, that the real costs are minimal. One of the best-known problems in this field is defining the optimal alignment for a road between two given points, a problem discussed in Section 6.3.2.

Another factor is of importance. The costs of a road can depend largely on the possible expansions which already are taken into account at the time of the original construction. E.g. sometimes very wide central reservations and shoulders are used to make later expansions possible at low cost. Moreover there is a difference between the cost for construction in one go and construction in stages (whether or not planned in advance). The cost related to this can be very high (demolition and reconstruction of structures and so on) especially when an expansion is not provided for in advance.

The investments in intersections are a very important share of the total amount of investments. However, these are even more difficult to put into cost functions, because many factors influence the shaping, and therefore the cost, of an intersection. There exists much literature (for instance A.A.S.H.O., 1965b, Beukers, 1967) and much practical experience with respect to the different possible shapings (cloverleaf, rotary, star etc.) and the choice between them.

Special structures are defined as those structures in the road for which high investments are necessary and which occur only incidentally and do not

belong to the 'normal' structures in the road or intersection. In the Netherlands special structures mean bridges, tunnels and aqueducts. The choice
between these alternatives on the basis of traffic flows on the road and on the
waterway, geographical situation and investment cost raises another (optimization) problem.

Also, in order to define the costs for the normal maintenance (repair of
the road-surface, maintenance of water-drainage systems and road-side
vegetation, repair or renewal of traffic signals, maintenance of parking-
places, running costs of the lighting and traffic-lights, grit-spreading etc.
in slippery conditions, keeping clean and so on) and renewal of the road-
surface many factors have to be considered, demanding cost models and
optimizations.

Moreover the costs for maintenance increase with increasing flow/
capacity ratio. This effect is neglected for reasons of simplicity and because
an optimal flow/capacity ratio is assumed, which is taken into account
when defining the costs for maintenance. When defining the optimal dimension this results in its being underestimated. But in view of the small share
of the costs of maintenance in the total costs, this error will not be serious.

9.5.2   Form of the Functions Used

On the basis of the considerations discussed above, Mulder (1971a)
defines construction and maintenance costs for different types of roads,
intersections and structures dependent on the number of (meeting) lanes.
The construction costs for intersections are assumed to depend on the total
number of lanes that meet in the intersection, independent of the way these
lanes are distributed over the legs of the intersection and how the main traffic
flows (through-going, left-turning, right-turning). This is, of course, a great
simplification. Next the cost amounts have been turned into linear functions
of the continuous variable c, which satisfy the necessary continuity conditions (see Section 10.1.3). In doing this, the fact that construction in stages
could be more expensive than construction in one go has been neglected.
This means that the costs for expansion from a 4-lane to a 6-lane road are
assumed to equal the difference between the construction costs for a 6-lane
and a 4-lane road.

The construction costs for road sections are represented by the following
relationship:

$$l(\text{ARINV} \cdot c + \text{BRINV}) \tag{9.5.1}$$

with

l—length of the road section
ARINV, BRINV—coefficients for the investments in road sections.

The construction costs of an intersection i are represented by the relationship:

$$\text{ARINT}_i\left(\sum_{\substack{k \\ (ik \in L)}} c_{ik} + \sum_{\substack{k \\ (ki \in L)}} c_{ki}\right) + \text{BRINT}_i \qquad (9.5.2)$$

with ARINT, BRINT–coefficients for the investments in intersections.

There are two ways of representing the construction costs of special structures. In the first case the special structure is indicated in the network as a node; in this case relationship (9.5.2) is applied, in which the 'intersection' usually then possesses two legs. In the second case the structure is represented as a link and then it is considered as a road section with special characteristics. In that case relationship (9.5.1) is applied.

For the cost of maintenance the following relationship is used:

$$l(\text{AMAINT} \cdot c + \text{BMAINT}) \qquad (9.5.3)$$

Concluding the 'investment costs' of the objective function for a link ij can be represented by $A_{ij}c + B_{ij}$ in which the coefficients A_{ij} and B_{ij} are composed as follows:

$$A_{ij} = \text{ARINV}_{ij}l_{ij} + \text{AMAINT}_{ij}l_{ij} + \text{ARINT}_i + \text{ARINT}_j$$

$$\begin{aligned} B_{ij} = {} & \text{BRINV}_{ij}l_{ij} + \text{BMAINT}_{ij}l_{ij} \\ & + \text{ARINT}_i\left(\sum_{\substack{k \neq j \\ (ik \in L)}} c_{ik} + \sum_{\substack{k \neq j \\ (ki \in L)}} c_{ki}\right) \\ & + \text{ARINT}_j\left(\sum_{\substack{l \neq i \\ (jl \in L)}} c_{jl} + \sum_{\substack{l \neq i \\ (lj \in L)}} c_{lj}\right) \\ & + \text{BRINT}_i + \text{BRINT}_j \end{aligned} \qquad (9.5.4)$$

N.B. If the node i represents a special structure the value $3c_{ij}$ is taken for

$$\left(\sum_{\substack{k \neq j \\ (ik \in L)}} c_{ik} + \sum_{\substack{k \neq j \\ (ki \in L)}} c_{ki}\right)$$

The constant term B_{ij} also contains the terms

$$\text{ARINT}_i \sum_{k \neq j} c_{ij}$$

and so on, so that part of the construction costs depends on the dimension of the other legs of the intersections at the ends of the related road section. So no special assignment of a part of the total costs to the related road section takes place, but the total costs are taken. Because the marginal construction costs are considered this is the correct procedure. In determining the total amount of investments for the whole road network, the investment costs of road sections and intersections are computed separately. It is necessary to make estimates for the dimensions of the other legs of the intersections. It is possible to take the values of a preceding step of the computation process for these. It is also possible to approximate these values better by applying, for instance, a Gauss–Seidel iteration process (see Hildebrand, 1956). In the network optimization of the Dutch Integral Transportation Study a Gauss–Seidel iteration process was applied, which was stopped after three steps (at a maximum).

9.5.3   Coefficients Used

The standard construction costs have been defined by Mulder (1971a) on the basis of data from the Rijkswaterstaat and the Provinciale Waterstaten (regional highway departments) (see also Rijkswaterstaat, 1969a and 1969b). The standard costs next have been turned into the functions discussed above in Section 9.5.2.

In the construction costs of road sections a distinction is made between five different road types:

circular roads;
roads in urban areas in the Randstad*;
roads in rural areas in the Randstad;
roads in urban areas in the remainder of the Netherlands;
roads in rural areas in the remainder of the Netherlands.

The functions for the construction costs are given in Figure 9.5.1. For some road sections, mainly in the Delta-area,† special investments functions are used which differ from the functions given.

In the construction of intersections a distinction is made between six types of intersections:

three-leg intersections in the Randstad;
three-leg intersections in the remainder of the Netherlands;
four-leg intersections in the Randstad;
four-leg intersections in the remainder of the Netherlands;
five-leg intersections in the Randstad;
five-leg intersections in the remainder of the Netherlands.

---

* The Ranstad (Rimcity) forms the most urbanized area of the Netherlands, including the towns Utrecht, Amsterdam, Zaandam, Haarlem, Leiden, The Hague, Delft, Vlaardingen, Schiedam, Rotterdam and Dordrecht.
† The Delta-area is the island-area in the south of Zuid-Holland and in Zeeland, where many important special structures, such as very long bridges and dikes, are to be built.

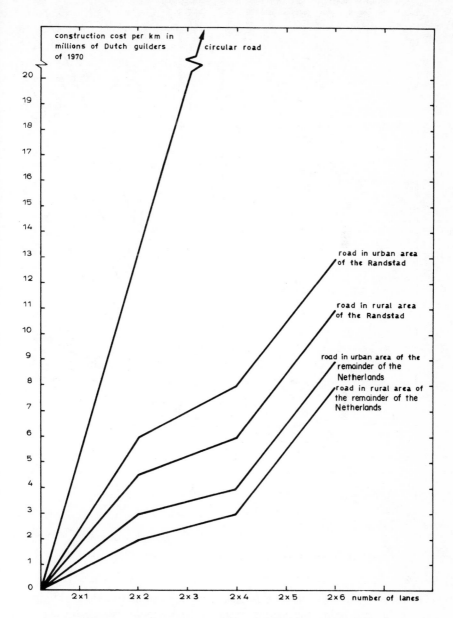

Figure 9.5.1   Construction cost for the different types of roads

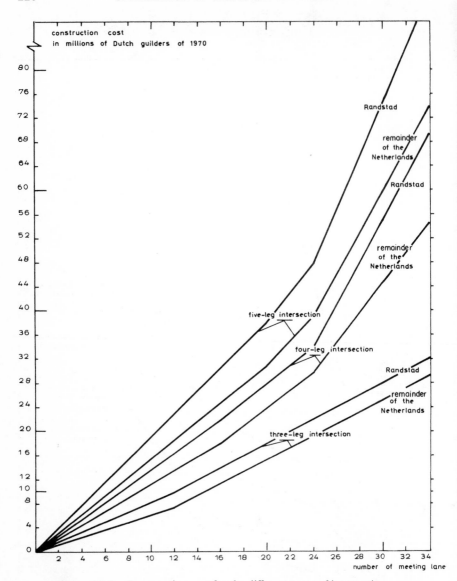

Figure 9.5.2   Construction cost for the different types of intersections

The different functions for the construction costs of intersections are given in Figure 9.5.2. Note that these investment curves do not possess the general shape described in Section 10.1.3 and shown in Figure 10.1.2 (the shape is even 'better', being totally convex). However, this does not cause any diffi-culty because for the coefficients of the construction costs of intersections

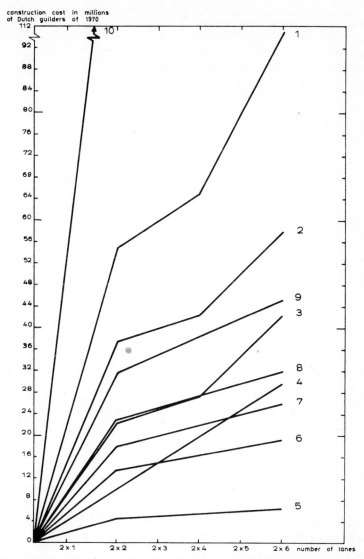

Figure 9.5.3   Construction cost for the different types of special structures.
Key:

1. tunnel under large waterway
2. tunnel under medium-sized waterway
3. tunnel under small waterway
4. aqueduct
5. bridge over canal
6. bridge over wide and well-used canal or small river
7. bridge over river
8. bridge over medium-sized river
9. bridge over large river
10. bridge in the Delta-area

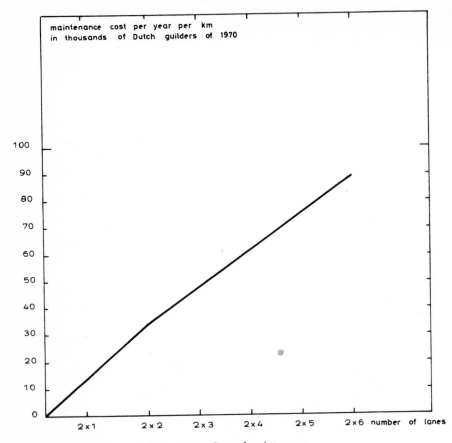

Figure 9.5.4   Cost of maintenance

for the computation of $F(x + \Delta x)$ the same values are taken as for the computation of $F(x)$. (In this way the functions acquire the shape described in Section 10.1.3).

In order to define the construction costs for special structures a distinction is made between ten types of special structures:

tunnels under large waterways;
tunnels under medium-sized waterways;
tunnels under small waterways;
aqueducts;
bridges over canals;
bridges over wide and well-used canals or small rivers;
bridges over rivers;
bridges over medium-sized rivers;
bridges over large rivers;
bridges in the Delta-area.

The different functions for the construction costs of special structures are given in Figure 9.5.3.

The costs for maintenance per year as a function of the number of lanes are given in Figure 9.5.4. These costs are computed for the whole period and discounted to present value (see Section 9.7).

9.6 COSTS TO THE ENVIRONMENT

Both the presence of a road and its use are a burden on the environment. That this is an important factor is proved by the frequent protests against the construction or expansion of roads. Such a disturbance of the present situation can affect the lives of many people in their homes, at work and/or in their recreation. Moreover it is becoming still clearer, that elements such as water, air and—for our problem the especially relevant element—space, are scarce, so that using or polluting these elements represents costs to the society.

Defining the environmental costs is very difficult. The difficulty arises in the registration and quantification of the environmental costs as well as in their social evaluation. Nowadays intensive research is being done in this field, but the problem has not been solved yet. Here we will mention some important aspects.

(a) *Damage to the beauty of towns and buildings*

Both for old and new cities whose buildings are of historical and/or architectural value and so on, the presence and/or use of roads can mean serious damage. This factor becomes even more serious when the function of the city is considered. It is obvious that we here encounter a most complex and difficult problem.

In the Dutch Integral Transportation Study only the interurban road network has been considered. Though the factors mentioned above also apply there, if only because roads and car traffic do not end at the borders of the town, they have not been included explicitly in the objective function. The factors did play a role in the definition of the costs for investments (see the high construction costs for circular roads, Figure 9.5.1) and in the definition of the possible network structure and the maximal dimensions of the different links.

(b) *Disturbance of the ecological balance*

The construction or expansion of a road may cause a serious disturbance of the existing ecological balance, which has been constituted over years, even centuries of plants, animals and people living and growing together. This is particular likely to cause serious damage when a new road is being constructed and whole areas can be cut through.

On the basis of these ecological factors a penalty has been defined for every link (Bos, 1971). This penalty serves as an indicator of the (relative) costs connected with the disturbance of the ecological balance caused by the construction or expansion of the related link. This penalty has yet to be converted into monetary cost, which would enable it to be added to the investments and other factors of the objective function. Because a wide road causes a more serious disturbance than a narrow one, a positive relationship between ecological costs and dimension seems plausible. A linear relationship, as between construction cost and dimension, however, would imply that a smaller dimension for the same traffic flow would be optimal. Indirectly this in its turn could cause a different choice of the routes in the network. However, it is doubtful whether this result is desired and, therefore, whether the relationships are not in reality more complicated.

Ignoring the possible relationships between ecological cost and such variables as flow and dimension it may be possible to relate the ecological cost directly to the value of the other elements of the objective function. In this case the positive relationship between ecological cost and dimension and traffic flow is retained and the ecological costs are involved during the network optimization. One possibility is to multiply the computed objective function directly by the penalty. This penalty might be dimensioned such that the (marginal) objective function of a road to which there are serious objections becomes three times as high. The idea behind this is that in a squared network a road three times the length should be used. The objective function for roads meeting with no objections should be kept constant.

Because of the uncertainty still existing about the monetary evaluation of the ecological costs it was ultimately decided not to include these factors in the objective function. As is known, ecological factors did play a role in defining the possible structure of the network and the restraints imposed upon the expansion of the different links.

(c) *Landscape aspects*

Related to the point mentioned above, the situation of the road in the terrain and the planting of trees, shrubs etc. play a role. This has consequences for driving comfort too (and so for the travel time evaluation) and for safety. It has been assumed that any extra costs related to all this, are included in the costs for construction and maintenance and so, implicitly, in the objective function.

(d) *Pollution of air and soil*

Exhaust gases contain many polluting and even dangerous components. So nowadays intensive research is being conducted into 'cleaner' fuel, 'cleaner' engines and the use in transport of other types of energy (electricity and so on).

Some of the costs arising from the pollution of air and soil depend wholly on the total number of driven kilometres and some of them also on the flow/capacity ratio of a road. For it can be stated that a high flow/capacity ratio gives rise to more pollution owing to the necessity of slowing down and pulling up when queues are formed.

(e) Noise

The noise caused by traffic on a highway can be reduced by good planting, digging in and so on. Where relevant, these costs have been assumed to be included in the costs for construction and maintenance. As with pollution, noise does depend partly on the flow/capacity ratio on the road.

Conclusion

The costs to the environment can form an important part of the total social costs. They can be important at the point of choosing the links to be constructed and used in the network and of defining the optimal flow/capacity ratio. However, the defining, the quantifying and the social evaluation of these types of costs are very difficult and are as yet in an early stage. Also, because of this fact, the costs to the environment have not been included explicitly in the objective function for the Dutch Integral Transportation Study. They did play a role in defining the constraints though.

9.7 THE DISCOUNT RATE

The discount rate serves to allow the comparison and addition of more or less continuously occurring costs, such as the users' and maintenance costs, with unique or periodically occurring investments. The total costs of the whole period are discounted to the present value in the starting year:

$$F = \int_{t_0}^{t_e} (F_t^{\text{TIME}} + F_t^{\text{RUN}} + F_t^{\text{ACC}} + F_t^{\text{MAINT}}) e^{-\pi t} \, dt + i \qquad (9.7.1)$$

with:

π—discount rate
$F_t^{\text{TIME}}, F_t^{\text{RUN}}, \ldots$—cost factors at year t
i—investments (in year t_0)
t_0—starting year of the related period
t_e—final year of the related period

In Section 10.5 the optimization of this type of function is discussed further. When the minimization of the objective function for one road forms a subproblem of a network optimization, it is impossible to say anything sensible about the course of the future flows and users' costs. So the costs changing over time have been assumed to stay constant and for the final year infinity has been chosen. Relationship (9.7.1) simplifies then to:

$$F = (1/\pi)(F^{\text{TIME}} + F^{\text{RUN}} + F^{\text{ACC}} + F^{\text{MAIN}}) + i \qquad (9.7.2)$$

The value of the discount rate indicates (to an extent) how heavy unique (large) expenses with long-lasting effects weigh against costs occurring annually. Many factors are relevant here, chief among which are the price for capital and the alternative possible uses of that (opportunity costs). The attitude towards risk is also involved: applying a high value for the discount rate the costs far in the future, which cannot be estimated very certainly, do not weigh heavily in the decision-making process. It is obvious that the discount rate may be differently defined if future costs or future benefits are concerned. For a further treatment of the discount rate, which is in fact a political factor, see for instance Baumol, 1968. For the Dutch Integral Transportation Study the value 10 per cent was chosen for the discount rate. This value is often applied for the cost-benefit studies of public investments.

9.8   The Objective Function and the Marginal Objective Function on a Road

The different components of the objective function and the different functions and coefficients in these functions have been given in the preceding sections. In Tables 9.8.1 and 9.8.2 the components of the two coefficients

Table 9.8.1   Indices for the elements of the coefficient D1

Year	CCR . KZERO	TZERO . KEXTRA	D1RUN'	D1ACC'	D1'
1980	50·0	21·3	18·2	10·5	100
1990	58·6	25·0	18·2	12·8	115
2000	65·0	27·8	18·2	15·7	126·5

Table 9.8.2   Indices for the elements of the coefficient D2

Year	TZERO . KZERO	D2RUN'	D2ACC'	D2'
1980	49·1	44·5	6·4	100
1990	57·6	44·5	7·9	110
2000	63·7	44·5	9·5	117·5

of the users' costs are given again, now expressed in indices, to show their relative weights.

Figure 9.8.1 illustrates the composition of the objective function for a certain road. In Figure 9.8.2 the objective function dependent on the traffic

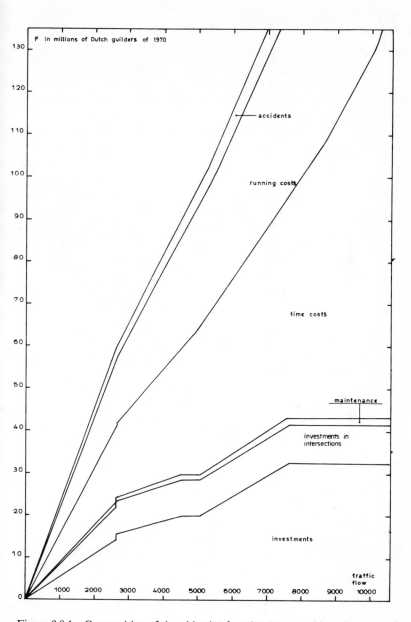

Figure 9.8.1   Composition of the objective function (new road in urban area of the Randstad with maximum dimensions of 12 lanes, length 5 kilometres with normal intersections at the ends)

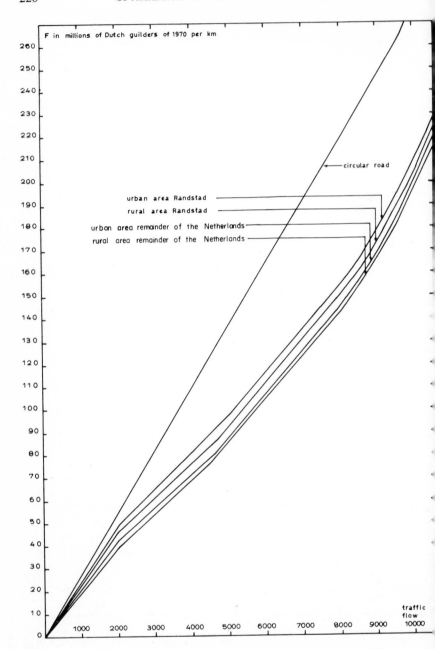

Figure 9.8.2   Objective function for the five main types of road

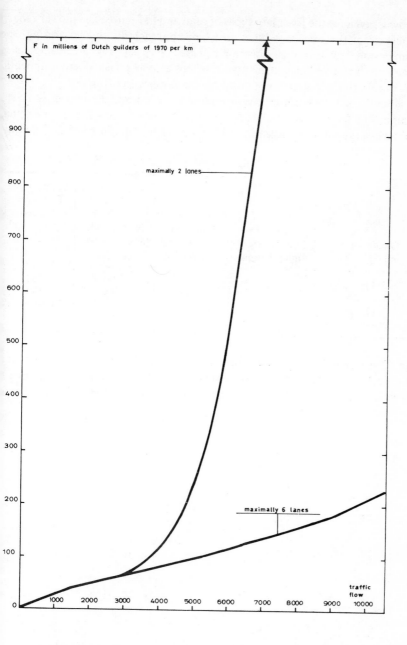

Figure 9.8.3   Objective function for dimension-constrained road

flow is given for the five main types of road and in Figure 9.8.3 the very sharp
increase of the objective function when approaching and 'exceeding' the
dimension constraint is illustrated. These very sharply increasing costs
may be considered as congestion costs, but also as physical, environmental
or prohibitively high investment costs (see also Section 10.1.4).

For the network optimization the marginal objective function is a very
important factor. As an illustration, some index values for the marginal
objective function are given in Table 9.8.3. This table shows that a 'detour'

Table 9.8.3   Index values for the marginal objective function

Type of road / Dimension interval	Circular road (5 km, normal intersections)	Urban area in Randstad (5 km, normal intersections)	Rural area in the remainder of the Netherlands (5 km, normal intersections)	Bridge in the Delta-area (special structure with 2 km normal road)
0–4 lanes	153	121	100	260
4–8 lanes	153	100	95	139
8–12 lanes	153	115	115	153
existing road	71	71	71	71

of 53 per cent to avoid a circular road is attractive. This means that, in a
network, a circular road is only used when absolutely necessary. The same is
true for the choice between an existing road and a new one. The new road

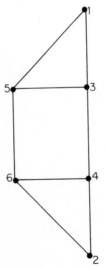

Figure 9.8.4   A new road in an existing network

must almost halve the distance to be attractive. Obviously this is only rarely the case in a network (see figure 9.8.4 : suppose 34 is a new road ; for the trip 34 the length of the road 3564 is about three times as much ; for the trip 12, use of 34 decreases the distance only by about 20 per cent).

REFERENCES

Adam, W. F. (1961). Safety aspects of motorway design, *Traffic Engineering and Control* (July).

A.N.W.B. (Royal Dutch Tourist Association), (1970). *Globale begroting voor een* .... A.N.W.B., 's-Gravenhage.

American Association of State Highway Officials (1965a). *Traffic Engineering Handbook, 3rd edition*, A.A.S.H.O., Washington D.C.

American Association of State Highway Officials (1965b). *Policy on Geometric Design of Rural Highways*, A.A.S.H.O., Washington D.C.

Baumol, W. J. (1968). On the social rate of discount. *The American Economic Review*, **58**, No. 4.

Beesley, M. C., and Evans, T. C. (1970). *The Costs and Benefits of Road Safety Measures*, Report of the Ninth Round Table on Transport Economics, European Conference of Ministers of Transport, Paris.

Behrendt, J., and Kloss, H. (1970). Congestie-onderzoek als bijdrage tot verkeersplanning en verkeersgeleiding, *Verkeerstechniek*, **21**, No. 12 (December); also in German (1970). *Strasse und Autobahn* (July).

Beukers, B. (1967). Vormgeving van ongelijkvloerse kruispunten, Preadvies Congresdag 1967: *Ongelijkvloerse Kruispunten in Auto(snel)wegen*, Vereniging Het Nederlandse Wegencongres, 's-Gravenhage.

Bos, P. (1971), *Rentabiliteit van Wegen en Niet-economische Criteria*, Deelrapport 19d, Integrale Verkeers- en Vervoerstudie door het Nederlands Economisch Instituut, Rotterdam.

Bruin, J. de (1971). *Ongevallen op Autowegen*. Nederlands Economische Instituut, Rotterdam.

Burgt, G. J. van der, Cate, A. J. ten, Hammendorp, H. J., and Kampen, L. T. B. van (1970). *Toekomstige Technische Ontwikkelingen in Verkeer en Vervoer*, Werkgroep wegvoertuigen, Deelrapport 7h, Integrale Verkeers- en Vervoerstudie door het Nederlands Economisch Instituut, Rotterdam.

Centraal Bureau voor de Statistiek (1967). *Het Bezit en Gebruik van Personenauto's*, 's-Gravenhage.

C.O.V.W. (Committee for the Promotion of Public Transport in the West of the Netherlands), (1966). *Data of the Home Survey on Magnetic Tapes*, Centrum van Vervoersplannen, Utrecht.

Donnea, F. X. de (1970). *Economische Waarde van Reistijdbesparingen in Nederland*, Deelrapport 19a, Integrale Verkeers- en Vervoerstudie door het Nederlands Economisch Instituut, Rotterdam.

Gwynn, D. W. (1967). Relationship of accident rates and accident involvements with hourly volumes, *Traffic Quarterly* (July).

Harrison, A. H., and Quarmby, D. A., (1970). *The Value of Time in Transport Planning : A Review*, Conférence Européenne des Ministres des Transport, Paris.

Henderson, J. M. and Quandt, R. E. (1958). *Microeconomic Theory : A Mathematical Approach*, McGraw Hill, New York.

Hendriks, A. J., and Vloemans, A. W. H. G. (1971). *Enige Sociaal-economische Aspecten van de Maatschappelijke Ontwikkeling in de Komende Decennia*, Deelrapport 11a, Integrale Verkeers- en Vervoerstudie door het Nederlands Economisch Instituut, Rotterdam.

Heijke, J. A. M. (1970a), *De Ontwikkeling van het Autoverkeer tot 2000; deel I: Het Personenautoverkeer*, Deelrapport 25a, Integrale Verkeers- en Vervoerstudie door het Nederlands Economisch Instituut, Rotterdam.

Heijke, J. A. M. (1970b). *Die Ontwikkeling van het Autoverkeer tot 2000; deel II: Het Vrachtautoverkeer; deel III: Het Totale Autoverkeer*, Deelrapport 25b, Integrale Verkeers- en Vervoerstudie door het Nederlands Economisch Instituut, Rotterdam.

Highway Research Board (1966). *Highway Capacity Manual, 1965*, Washington D.C.

Hildebrand, F. B. (1956). *Introduction to Numerical Analysis*, New York.

Mulder, T. (1971a). *Investerings- en Onderhoudskosten van de Weg*, Deelrapport 19b, Integrale Verkeers- en Vervoerstudie door het Nederlands Economisch Instituut, Rotterdam.

Mulder, T. (1971b). *De Relaties tusten Snelheid, Intensiteit en Capaciteit op Wegen voor Interstedelijk Verkeer*, Deelrapport 1a, Integrale Verkeers- en Vervoerstudie door het Nederlands Economisch Instituut, Rotterdam.

Nederlands Vervoerswetenschappelijk Instituut (1950). *De Schade Veroorzaakt door Wegverkeersongevallen*, Rapport uitgebracht aan de Minister van Verkeer en Waterstaat, Rotterdam.

Oort, C. J. (1969). Evaluation of travelling time, *Journal of Transport Economics and Policy* (September).

Pfund and Koenmerer (1970). The influence of certain roadway and traffic characteristics of rural roads on the level of service and safety, *IX International Study Week in Traffic Safety Engineering*.

Rijkswaterstaat (1969a). *Gemiddelde Kosten per km Autosnelweg in de Randstad en Overig Nederland*, Rijkswaterstaat, 's-Gravenhage.

Rijkswaterstaat (1969b). *Benodigde Middelen voor Onderhoud en Verbetering van Rijkswegen en Oeververbindingen*, Rijkswaterstaat, 's-Gravenhage.

Schlums, J. (1967). Probleme des Dimensionierung von Anlagen des Strassenverkehrs, *Strasse und Autobahn* (August and September).

Soberman, R. M., and Clark, G. A. (1970). The role of vehicle operating costs in highway planning. Paper presented at *The Annual Convention of the Canadian Good Roads Association in Conjunction with the VI World Highway Conference*, Montreal (October 4–10).

Stichting Wetenschappelijk Onderzoek Verkeersveiligheid (1963). *De Schade van de Gemeenschap Veroorzaakt door de Verkeersongevallen op de Openbare Weg in 1962*, 's-Gravenhage.

Winch, D. M. (1963). *The Economics of Highway Planning*, Toronto.

Winfrey, R. (1969). *Economic Analysis for Highways*, International Textbook Co., Scranton Pa.

10

The Optimal Number of Lanes for a Given Traffic Flow

In this chapter the subproblem of the transport network optimization problem is dealt with, namely the definition of the optimal dimension for a given traffic flow. In Section 10.1 the problem is discussed in general terms, using the general shape of the objective function of Chapter 9. Working with the elaborated objective function of Chapter 9 in Section 10.2 some sensitivity analyses of the results are discussed. The final determination of the optimal number of lanes for a given traffic flow is treated in Section 10.3. Finally in the Sections 10.4 and 10.5 respectively some remarks are made about the possible transitions from a continuous dimension to a discrete number of lanes and about the optimization over time.

10.1 SHAPE OF THE OBJECTIVE FUNCTION AND THE MATHEMATICAL SOLUTION OF THE SUBPROBLEM

10.1.1 Statement of the Problem

For the subproblem the total social costs as discussed in the preceding chapter for one link form the objective function. Omitting the subscripts ij the objective function for one link has the following general form:

$$F = xt(x, c) + i(c) \qquad (10.1.1)$$

with:

    c—dimension of the road
    i—investments and maintenance of the road
    x—traffic flow
    t—users' costs

The constraints are formed by the given traffic flow and the minimum and maximum dimensions:

$$x = x^0$$

$$c^{min} \leqslant c \leqslant c^{max}$$

(10.1.2)

with

c^{min}—existing dimension
c^{max}—maximum dimension
x^0—given traffic flow

10.1.2   Solution Methods for the Minimization Problem

10.1.2.1   Exhaustive Enumeration

If a road is supposed to be constructed or expanded with only a restricted number of possibilities for the number of lanes (say 2-, 4-, 6- and 8-lane roads), the simplest minimization method is exhaustive enumeration. In this method the value for the objective function is computed for all feasible values for the decision variable(s) and then that case is selected in which the objective function has the lowest value.

10.1.2.2   Search Methods

When the number of possibilities becomes larger the exhaustive enumeration becomes less efficient. In our case the number of possible solutions can become larger, because for instance 10-lane and 3-lane roads are also considered to be reasonable, but also because roads with a more or less continuous number of lanes are considered to be reasonable. In that case a 3·1-lane road may be interpreted as a 2-lane road, that needs to be expanded to a 4-lane road later than say a 3·4-lane road.

When a search method is used, the objective function is also computed for a number of feasible values for the decision variable(s). However, a new value for the decision variable, for which the objective function is computed, is now defined on the basis of earlier results. Search methods can be applied

Table 10.1.1   Search method

Computation	Comparison	Conclusion
$F(c_1)$		
$F(c_2)$	$F(c_2) < F(c_1)$	$c^* > c_1$
$F(c_3)$	$F(c_3) > F(c_2)$	$c_1 < c^* < c_3$
$F(c_4)$	$F(c_2) < F(c_4) < F(c_3)$	$c_1 < c^* < c_4$
$F(c_5)$	$F(c_1) > F(c_5) > F(c_2)$	$c_5 < c^* < c_4$

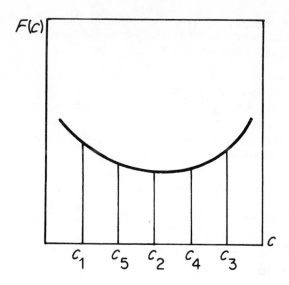

Figure 10.1.1 Search method

most successfully, for convex (or concave in the case of maximization) functions. As is known, the objective function for the transport network optimization is convex with respect to the dimension (infinite users' cost for dimension zero and infinite investments for dimension infinity) so a search method can be applied successfully here.

The operation of a search method is illustrated with a small example in Figure 10.1.1 and Table 10.1.1. In the way that is shown in Table 10.1.1 and Figure 10.1.1, a continually diminishing interval can be computed in which the optimal solution c^* is to be found. There exist rules for the choice of the next values of c, in such a way that the number of tests is as small as possible (Wilde, 1964).

10.1.2.3 Solution by Differentiation

When the variable c is continuous and when the function $F(c)$ is differentiable and convex, the minimum is reached when the differential quotient of $F(c)$ equals zero. The optimal value c^* is computed, then, by solving the equation:

$$\frac{\mathrm{d}F}{\mathrm{d}c} = 0 \tag{10.1.3}$$

This method is used for the solution of the subproblems of the network optimization problem for the Dutch Integral Transportation Study.

10.1.3   Shape of the Functions Used and Mathematical Relationships between Dimension and Traffic Flow and between Objective Function and Traffic Flow

In this subsection we will discuss the mathematical shape of the functions actually used and in correspondence with this the mathematical relationships between the optimal dimension and the traffic flow and between the optimal value for the objective function and the traffic flow.

The users' costs are represented by the following function of the traffic flow and the dimension:

$$z = x \left\{ D2 + D1 \left(\frac{x}{c} \right)^{CCRE} \right\} \tag{10.1.4}$$

with $D1$, $D2$, CCRE—coefficients: The value of the exponent CCRE in this function is always larger than one.

The costs for investments and maintenance are assumed linear dependent on the dimensions:

$$i = Ac + B \tag{10.1.5}$$

with A, B—coefficients

Generally the coefficients A and B have different values for maximal three intervals of the dimension. Assuming in this subsection $c^{min} = 0$ and c^{max} is large enough, we get the following values for A and B:

$$A1, B1 \quad 0 \leqslant c \leqslant c^{12}$$

$$A2, B2 \quad \text{for } c^{12} \leqslant c \leqslant c^{23}$$

$$A3, B3 \quad \text{for } c^{23} \leqslant c \leqslant c^{max}$$

where the function $i(c)$ is continuous and equals zero when the dimension equals zero:

$$A1 \cdot 0 + B1 = 0 \longrightarrow B1 = 0 \tag{10.1.6}$$

$$A1 \cdot c^{12} + B1 = A2 \cdot c^{12} + B2 \longrightarrow c^{12} = \frac{B2 - B1}{A1 - A2} \tag{10.1.7}$$

$$A2 \cdot c^{23} + B2 = A3 \cdot c^{23} + B3 \longrightarrow c^{23} = \frac{B3 - B2}{A2 - A3} \tag{10.1.8}$$

Moreover, the following relationships hold for the coefficients A:

$$A1 \geqslant A2 \quad \text{and} \quad A2 \leqslant A3 \tag{10.1.9}$$

The general investment function is shown in Figure 10.1.2. This threefold investment function and the constraint(s) for the dimension give rise to

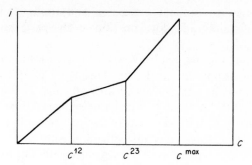

Figure 10.1.2   Investments as a function of the
dimension (general shape)

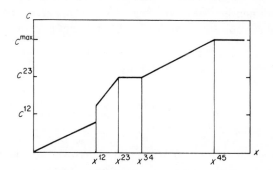

Figure 10.1.3   Optimal dimension as a function
of the traffic flow (general shape)

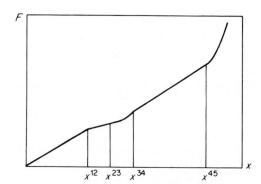

Figure 10.1.4   Objective function as a function
of the traffic flow (general shape)

relationships between optimal dimension and traffic flow and between objective function and traffic flow consisting of five parts (see Figures 10.1.3 and 10.1.4).

In the first part the marginal investment costs ($A1$) are fairly high so a fairly small dimension is optimal then. Through a jump in the dimension (at $x = x^{12}$) the second part is reached where the marginal investment costs are smaller and so the increments of the dimension are larger. The dimension increases till the third, and steep, part of the investment function is reached. Now the dimension stays constant with increasing traffic flow until the, now more expensive, expansion ($A3$) is sensible again. The road is expanded, till the upper bound for the dimension is reached.

The course of the objective function as a function of the traffic flow follows immediately—linear increment where the dimension increases linearly with the traffic flow; increment according to a function of the power $CCRE + 1$, where the dimension stays constant. This is apparent from a further working out of the objective function for the different intervals.

Interval 1: $0 \leqslant x \leqslant x^{12}$

The objective function is stated as follows:

$$ F = x \left\{ D2 + D1 \left(\frac{x}{c} \right)^{CCRE} \right\} + A1c + B1 $$

Putting the differential quotient of F equal to zero and solving the resulting equation gives the relationship between optimal dimension and traffic flow:

$$ c = x \sqrt[CCRE+1]{\frac{D1 \cdot CCRE}{A1}} \tag{10.1.10} $$

Substitution of c into the objective function and some working out gives the relationship between objective function and traffic flow:

$$ F^{min}(x) = x \left\{ D2 + D1^{\frac{1}{1+CCRE}} A1^{\frac{CCRE}{1+CCRE}} \left(CCRE^{\frac{-CCRE}{1+CCRE}} + CCRE^{\frac{1}{1+CCRE}} \right) \right\} + B1 $$

$$ \tag{10.1.11} $$

So we see that $F^{min}(x)$ is a linear function and goes through the origin, because $B1 = 0$ [equation (10.1.6)].

The value of x^{12} follows from the equation:

$$ F^{min}{}_1(x^{12}) = F^{min}{}_2(x^{12}) $$

in which $F^{\min}{}_1$ is the objective function computed with $A1$ and $B1$ and $F^{\min}{}_2$ is the one computed with $A2$ and $B2$. Working out gives the following value for x^{12}:

$$x^{12} = \frac{B2 - B1}{D1^{\frac{1}{1+\mathrm{CCRE}}}\left(A1^{\frac{\mathrm{CCRE}}{1+\mathrm{CCRE}}} - A2^{\frac{\mathrm{CCRE}}{1+\mathrm{CCRE}}}\right)\left(\mathrm{CCRE}^{\frac{-\mathrm{CCRE}}{1+\mathrm{CCRE}}} + \mathrm{CCRE}^{\frac{1}{1+\mathrm{CCRE}}}\right)}$$

(10.1.12)

It can be shown that this value is indeed correct, that means:

$$c(x) \leqslant c^{12} \quad \text{for } x < x^{12}$$
$$c(x) \geqslant c^{12} \quad \text{for } x > x^{12}$$

(10.1.13)

To prove statement (10.1.13) it is proved that:

$$c_1(x^{12}) \leqslant c^{12} \leqslant c_2(x^{12})$$

(10.1.14)

with:

$$c_1(x^{12}) = x^{12}\left(\frac{D1 \cdot \mathrm{CCRE}}{A1}\right)^{\frac{1}{1+\mathrm{CCRE}}}$$

and:

$$c_2(x^{12}) = x^{12}\left(\frac{D1 \cdot \mathrm{CCRE}}{A2}\right)^{\frac{1}{1+\mathrm{CCRE}}}$$

To prove relationship (10.1.14) the equations (10.1.12) for x^{12} and (10.1.7) for c^{12} are substituted into the inequality (10.1.14). Some working out shows that the first part of the inequality is equivalent to:

$$A1^{\frac{1}{1+\mathrm{CCRE}}}A2^{\frac{\mathrm{CCRE}}{1+\mathrm{CCRE}}} \leqslant \frac{A1 + \mathrm{CCRE} \cdot A2}{1 + \mathrm{CCRE}}$$

(10.1.15)

and the second part of the inequality is equivalent to:

$$A1^{\frac{\mathrm{CCRE}}{1+\mathrm{CCRE}}}A2^{\frac{1}{1+\mathrm{CCRE}}} \leqslant \frac{\mathrm{CCRE} \cdot A1 + A2}{1 + \mathrm{CCRE}}$$

(10.1.16)

The inequalities (10.1.15) and (10.1.16) are the inequalities of Cauchy and they are in fact true (Hardy, Littlewood and Polya, 1934). So the statement (10.1.14) is proved. (N.B. When $x = x^{12}$ there are thus two solutions for the dimension with the same value for the objective function; so both solutions are correct.)

Interval 2: $x^{12} \leqslant x \leqslant x^{23}$

The shape of the functions for the optimal dimension and the objective function are similar to those for the first interval:

$$c = x^{\text{CCRE}+1}\sqrt{\dfrac{D1.\text{CCRE}}{A2}} \tag{10.1.17}$$

$$F^{\min}(x) = x\left\{D2 + D1^{\frac{1}{1+\text{CCRE}}}A2^{\frac{\text{CCRE}}{1+\text{CCRE}}}\left(\text{CCRE}^{\frac{-\text{CCRE}}{1+\text{CCRE}}} + \text{CCRE}^{\frac{1}{1+\text{CCRE}}}\right)\right\} + B2 \tag{10.1.18}$$

The value of x^{23} follows from:

$$c_2(x^{23}) = c^{23}$$

so:

$$x^{23} = c^{23}\,{}^{\text{CCRE}+1}\sqrt{\dfrac{A2}{D1.\text{CCRE}}} \tag{10.1.19}$$

Interval 3: $x^{23} \leqslant x \leqslant x^{34}$

In this third interval the dimension is constant and equals c^{23}:

$$c = c^{23} \tag{10.1.20}$$

The objective function is now a function of the power $\text{CCRE} + 1$:

$$F^{\min}(x) = x\left\{D2 + D1\left(\dfrac{x}{c^{23}}\right)^{\text{CCRE}}\right\} + A2c^{23} + B2 \tag{10.1.21}$$

The value of x^{34} follows from:

$$c_3(x^{34}) = c^{23}$$

so:

$$x^{34} = c^{23}\,{}^{\text{CCRE}+1}\sqrt{\dfrac{A3}{D1.\text{CCRE}}} \tag{10.1.22}$$

To show that c^{23} is indeed the optimal solution for the third interval of x we look at the derivatives of $F(x, c)$ for $x^{23} \leqslant x \leqslant x^{34}$. Left of c^{23} $F(x, c)$ is replaced by $F_2(x, c)$ and right of c^{23} $F(x, c)$ is replaced by $F_3(x, c)$.

Taking the left derivative† of $F(x, c)$ with respect to c we see that this left derivative is negative for $x > x^{23}$. So:

$$c^* \geqslant c^{23} \text{ for } x > x^{23} \tag{10.1.23}$$

---

† The left derivative is the following limit:

$$\lim_{\Delta c \uparrow 0} \dfrac{F(c + \Delta c) - F(c)}{\Delta c}$$

Taking the right derivative of $F(x, c)$ with respect to c we see that this right derivative is positive for $x < x^{34}$. So:

$$c^* \leqslant c^{23} \quad \text{for } x < x^{34} \tag{10.1.24}$$

Combination of relationships (10.1.23) and (10.1.24) gives:

$$c^* = c^{23} \quad \text{for } x^{23} < x < x^{34}$$

Interval 4: $\quad x^{34} \leqslant x \leqslant x^{45}$

The fourth interval is again a 'normal' interval with:

$$c = x \sqrt[\text{CCRE}+1]{\frac{D1 \cdot \text{CCRE}}{A3}} \tag{10.1.25}$$

and:

$$F^{\min}(x) = x \left\{ D2 + D1^{\frac{1}{1+\text{CCRE}}} A3^{\frac{\text{CCRE}}{1+\text{CCRE}}} \left(\text{CCRE}^{\frac{-\text{CCRE}}{1+\text{CCRE}}} + \text{CCRE}^{\frac{1}{1+\text{CCRE}}} \right) \right\} + B3 \tag{10.1.26}$$

The value for x^{45} follows from the upper bound for the dimension and is given by the relationship:

$$x^{45} = c^{\max} \sqrt[\text{CCRE}+1]{\frac{A3}{D1 \cdot \text{CCRE}}} \tag{10.1.27}$$

Interval 5: $\quad x_{45} \leqslant x$

In the fifth and last interval the optimal dimension equals the maximum value for the dimension and the objective function is again a function of the power $\text{CCRE} + 1$ of the traffic flow:

$$c = c^{\max} \tag{10.1.28}$$

$$F^{\min}(x) = x \left\{ D2 + D1 \left(\frac{x}{c^{\max}} \right)^{\text{CCRE}} \right\} + A3 c^{\max} + B3 \tag{10.1.29}$$

10.1.4 Maximum and Minimum Value for the Dimension

As stated in relationship (10.1.2) the dimension of a road is constrained by a lower bound $c^{\min}$ and an upper bound $c^{\max}$:

$$c^{\min} \leqslant c \leqslant c^{\max} \tag{10.1.30}$$

The lower bound is generally formed by the existing dimension and is always greater than or equal to zero. There is an upper bound defined for every road. The meaning of the upper bound is that it is physically impossible, or not desired with respect to the damage to the environment or the existing building, or with respect to the operation of the traffic or otherwise, to expand the road beyond this upper bound. Or it may be possible that the costs of investing are prohibitively high. Instead of defining all different functions for these cost factors, as would be correct theoretically, for the sake of simplicity a prohibition is put on further expansion.

The effect of $c^{\max}$ has been shown above in Section 10.1.3. Of course it is not necessary for this maximum dimension to lie at the end of the third interval of the investment function. The effect of the minimum dimension is obvious:

if

$$x \leqslant c^{\min} \sqrt[\text{CCRE}+1]{\frac{A_{c\min}}{D1 \cdot \text{CCRE}}}$$

then:

$$c = c^{\min} \tag{10.1.31}$$

$$F^{\min}(x) = x\left\{D2 + D1\left(\frac{x}{c^{\min}}\right)^{\text{CCRE}}\right\} + \text{AMAINT}_{c\min}c^{\min} + \text{BMAINT}_{c\min}$$

$$\tag{10.1.32}$$

with:

$A_{c\min}$ —coefficient A for the interval of $c^{\min}$
$\text{AMAINT}_{c\min}$—coefficient A for the costs of maintenance for the interval of $c^{\min}$
$\text{BMAINT}_{c\min}$—coefficient B for the costs of maintenance for the interval of $c^{\min}$

and if:

$$x > c^{\min} \sqrt[\text{CCRE}+1]{\frac{A_{c\min}}{D1 \cdot \text{CCRE}}}$$

then:

$$c = x \sqrt[\text{CCRE}+1]{\frac{D1 \cdot \text{CCRE}}{A_c}} \tag{10.1.33}$$

or otherwise, depending on the interval, according to equations (10.1.20) or 10.1.28), and:

$$c^{min}(x) = x\left\{D2 + D1\left(\frac{x}{c}\right)^{CCRE}\right\} + AMAINT_c c + BMAINT_c + ARINV_c c$$
$$+ BRINV_c - ARINV_{cmin}c^{min} - BRINVc^{min} \qquad (10.1.34)$$

or otherwise, depending on the interval, according to equations (10.1.21) or (10.1.29).
with:

A_c —coefficient A for the interval of c
$ARINV_c$ —coefficient A for the investment costs for the interval of c
$ARINV_{cmin}$—coefficient A for the investment costs for the interval of c^{min}
$BRINV_c$ —coefficient B for the investment costs for the interval of c
$BRINV_{cmin}$—coefficient B for the investment costs for the interval of c^{min}

N.B. To take into account the different possible intervals of x and their consequences, in the computer program c is computed first ignoring the constraints and after that it is inspected to see if $c < c^{min}$ or $c > c^{max}$.)

10.2 SOME SENSITIVITY ANALYSIS

In the preceding chapter the values for the different cost factors depending on the traffic flow and the number of lanes have been discussed. It has become clear that in defining the cost functions many approximations and many assumptions had to be made where it was not possible to be definite. Moreover, some factors could be used as decision variables, for example the distribution of the traffic over time and the discount rate. For all these reasons it is extremely important to know the effect of alterations in the assumptions. We will give this some attention in this subsection, without presenting a full sensitivity analysis.

We will discuss the sensitivity of the optimal relationship between the traffic flow and the dimension and of the objective function and the marginal objective function.

10.2.1 Sensitivity of the Optimal Relationship between the Traffic Flow and the Number of Lanes

On a 'normal' interval of the objective function, that is one that is beyond the influence of the dimension constraints and/or changes in the slope of the investment curve, the optimal dimension for a given traffic flow is given by relationship (10.1.10).

$$c = x^{\text{CCRE}+1} \sqrt{\frac{D1 \cdot \text{CCRE}}{A}} \qquad (10.1.10$$

We will discuss the sensitivity of this relationship in four parts:

(a) the composition of the coefficient $D1$
(b) the sensitivity for the quotient $D1/A$;
(c) the sensitivity for the exponent CCRE;
(d) the sensitivity for the functional relationship of the users' costs.

(a) *The composition of the coefficient* $D1$

As is known, the coefficient A forms the marginal cost of investment and the coefficient $D1$ is composed of the coefficients related to the parts of the users' costs depending on the flow/capacity ratio. To get a better idea of the relative weight of the different parts constituting the coefficient $D1$ and so of the sensitivity of the optimal dimension for these different parts we will have another look at this coefficient $D1$. It is composed of the travel time running and accident costs in the following way:

$$D1 = \sum_h (\text{CCR} \cdot \text{KZERO} + \text{TZERO} \cdot \text{KEXTRA} + \text{D1RUN}'$$
$$+ \text{D1ACC}')\left(\frac{x_h}{x_b}\right)^{\text{CCRE}+1} \qquad (10.2.$$

The values for these coefficients have been treated in the preceding chapter. Their relative weights and their relative influence on the optimal flow capacity-ratio is shown in Table 9.8.1.

(b) *The sensitivity for the quotient* $D1/A$

From relationship (10.1.10) it follows that the optimal number of lanes for a given traffic flow increases with increasing users' costs and thus with increasing travel time costs, running costs, accidental costs, value of travel time and/or real income and/or with decreasing marginal investment and maintenance costs and/or discount rate. In Table 10.2.1 this sensitivity for the quotient $D1/A$ is shown for different values for the exponent CCRE. In this table the optimal flow/capacity ratio (index) is given as a function of the quotient $A/D1$. Furthermore in the table some possible causes for the changes in $A/D1$ are given to get some idea of the influence of the different parameters.

It is clearly shown in Table 10.2.1 that the sensitivity of the optimal flow/capacity ratio for the quotient $A/D1$ decreases with increasing value for

Possible causes for the variation

Year	Value of travel time (in Dutch guilders per hour)	Discount rate (in %)	Road type	A/D1 (index)	Optimal flow/capacity ratio (indices)				
					CCRE = 3	CCRE = 4	CCRE = 5	CCRE = 6	CCRE = 7
	50·20	1		10	56	63	68	72	75
	19·20	2·5		25	71	76	79	82	84
	8·75	5		50	84	87	89	91	92
	5·90	7	Rural area remainder of the Netherlands, 4–8 lanes	70	91	93	94	95	96
2000	5·42	7·5		75	93	94	95	96	96
	5·15	7·8		78	94	95	96	97	97
1990	4·95	8		80	95	96	96	97	97
	4·46	8·7		87	97	97	98	98	98
1980	4·25	9		90	97	98	98	99	99
	3·68	10	Urban area Randstad, 4 8 lanes; rural area remainder of the Netherlands, starting investments	100	100	100	100	100	100
	3·22	11		110	102	102	102	101	101
	2·82	12		120	105	104	103	103	102
	1·98	15		150	111	109	107	106	105
		17	Normal road with more than 8 lanes	170	114	111	109	108	107
		19	Urban area Randstad. starting investments	190	117	114	111	110	108
		20	Bridge in Delta-area, 4–8 lanes	200	119	115	112	110	109
		29	Circular road: bridge in Delta-area. 8–12 lanes	290	130	124	119	116	114
		36		360	138	129	124	120	117
		50	Bridge in Delta-area, starting investments	500	150	138	131	126	122
		100		1·000	178	158	147	139	133

the exponent CCRE. A high value for CCRE implies an initial slow increase
of users' costs with increasing flow/capacity ratio but then a sudden sharp
increase. At that moment investing is always optimal almost independent of
the construction cost and/or discount rate.

For the rest, the optimal flow/capacity ratio turns out to be rather in
dependent of factors such as the value of travel time and the discount rate
An increase of the discount rate from 10 to 15 per cent means an increase in
the optimal flow/capacity ratio of 7 per cent with CCRE = 5 and of 5 per cen
with CCRE = 7; doubling the value of travel time causes a decrease of the
optimal flow/capacity ratio of 10 per cent and 6 per cent respectively. On the
other hand, the differences in the marginal investment costs of the different
road types are very large indeed. If, to get an idea, the optimal flow/capacity
ratio of an urban road in the Ranstad of 4 to 8 lanes is supposed to be 1,200
p.c.u. per lane for the average evening peak hour, we get the following value
for the optimal flow/capacity ratio of a circular road for CCRE from 3 to '
respectively: 1,660, 1,550, 1,490, 1,440 and 1,400 p.c.u. per lane for the
average evening peak hour.

(c) *The sensitivity for the exponent* CCRE

The effect of the coefficient CCRE, the exponent in the travel time and
other users' cost functions on the optimal flow/capacity ratio is somewhat
more complicated than that of the coefficients $D1$ and A. In the above it ha
already been shown that differences in $D1$ and A have less influence as the
value of CCRE increases. This appears also directly from relationshi
(10.1.10). Furthermore the coefficient CCRE influences the coefficient $D1$, a
is shown in relationship (10.2.1). From this relationship it is seen that thi
influence depends on the supposed distributions of the traffic over time.

To inspect the sensitivity for exponent CCRE at different assumptions fc
the distribution over time of the traffic (or, formulated the other way round
the sensitivity for the distributions over time at different values for CCRE
the optimal flow/capacity ratio has been computed for different distribution
over time and for different values for CCRE. For the possible distribution
the distributions for the traffic on work days for the different years, mentione
in Section 9.2.3, composed by Hendriks and Vloemans (1970) (Figures 9.2.
and 9.2.8) have been used. The most important characteristics from thes
distributions are:

> *1980 :* pronounced peak from 5 to 6 p.m.;
> *1990 :* peaks from 7 to 9 a.m. and from 2 to 4 p.m. and from 5 to 6 p.m
> *2000 :* high peak from 2 to 4 p.m.

For the flow on the average evening peak hour for 1980 the flow from 5 t
6 p.m. has been taken and for 1990 and 2000 the flow from 3 to 4 p.m., all on
work day in October.

First our attention will be focused on the coefficient $D1$. The higher the exponent CCRE, the heavier the peak hours weigh and the lighter the valley hours. If the average annual hourly flow is taken as basis (see Section 9.2.3), $D1$ increases sharply with increasing CCRE. Furthermore, a less smooth distribution also causes a higher value for $D1$, especially at high values for CCRE. If $D1$ is computed on the basis of the average evening peak hour, $D1$ decreases with increasing CCRE and less smooth distributions over time. For a diminishing number of hours makes an important contribution to the total costs. These trends are shown in Table 10.2.2.

Table 10.2.2   Sensitivity of $D1$ for the distributions over time of the traffic and for the exponent CCRE

Distribution for the year	Exponent CCRE	$D1$ (index) on the basis of the average annual hourly flow	$D1$ (index) on the basis of the flow on the average evening peak hour
1980	3	28	165
	4	51	124
	5	100	100
	6	196	81
	7	446	76
1990	3	28	373
	4	50	332
	5	92	308
	6	174	292
	7	331	280
2000	3	37	211
	4	78	184
	5	171	165
	6	382	151
	7	881	143

Table 10.2.2 shows clearly the high sensitivity of the coefficient $D1$ for distributions over time of the traffic and for the exponent CCRE. Of course it can be easily understood that it makes some difference if two high peak hours exist, as has been supposed for the distribution of the year 2000, one high peak hour, as for 1980, or a large number of lower peaks, as for 1990. The higher the exponent CCRE, the higher the relative weight of the peaks, as can be seen also clearly in Table 10.2.2.

The ultimate effect of CCRE on the optimal number of lanes for a given traffic flow is of course more interesting. Based on the average annual hourly flow, it is obvious that at a flat distribution over the day, as assumed for

1990, only a few lanes are necessary; if only one peak hour exists, as for 1980, the need is not too large either, but two high peak hours as for 2000 cause a large need for infrastructure. The lower the value of CCRE, the greater the number of hours that contribute substantially to the total costs, so the more lanes are necessary (see Figure 9.2.5; the speed is lower for lower values for CCRE in the interval from 800 to 1,800 p.c.u. per hour per lane).

The reverse effect, *viz.* that the highest congestion costs on the peak hour weigh heavier as CCRE increases does not outweigh this effect. These trends are shown in Table 10.2.3.

Table 10.2.3 Sensitivity of the optimal number of lanes for a given annual hourly flow for the distribution over time of the traffic and for the exponent CCRE

Exponent CCRE	Number of lanes for a given annual hourly flow (index)		
	1980	1990	2000
3	127	127	133
4	110	109	120
5	100	99	110
6	93	92	103
7	89	86	98

It is interesting to represent these figures in another way, namely as the optimal flow/capacity ratio. This is done for the distribution of the year 1980, both in indices and in p.c.u. per lane (with 1,350 p.c.u. as reference) in Table 10.2.4.

Table 10.2.4   Sensitivity of the optimal flow/capacity ratio for the exponent CCRE (for the year 1980)

Exponent CCRE	Optimal flow/capacity ratio (index)	Optimal flow/capacity ratio in p.c.u. per hour per lane
3	79	1,070
4	91	1,230
5	100	1,350
6	107	1,440
7	112	1,510

Comparison of the sensitivity of the optimal number of lanes for the exponent CCRE with that for factors such as construction costs, discount rate, travel time evaluation and so on shows an alarmingly greater effect of CCRE.

(d) *The sensitivity for the functional relationship of the users' costs*

For all parts of the users' costs, the same functional relationship has been used. Of course, many objections can be made against this.

As an example of the sensitivity of the optimal flow/capacity ratio for the form of the function, one different form of the objective function will be dealt with. The following form for the objective function is chosen:

$$F = x \left\{ D1 \left(\frac{x}{c}\right)^{CCRE1} + D3 \left(\frac{x}{c}\right)^{CCRE3} + D2 \right\} + Ac + B \qquad (10.2.2)$$

Taking the value 7 for CCRE1 and the value 3 for CCRE3 the optimal flow/capacity ratio is given by:

$$\frac{x}{c} = \left\{ \frac{(CCRE3^2 . D3^2 + 4CCRE1 . D1 . A)^{1/2} - CCRE3 . D3}{2CCRE1 . D1} \right\}^{1/4} \qquad (10.2.3)$$

Representing the travel time costs by the 7th power function and the running and accident costs by the 3rd power function the value 275·24 must be chosen for $D1$ and the value 41·74 for $D3$, to obtain cost functions, defined in the same way as indicated in Sections 9.2, 9.3 and 9.4. The values for the optimal flow/capacity ratio of a road in an urban area of the Randstad with a length of 5 kilometres and normal intersections at the ends are given in Table 10.2.5 together with the values obtained applying the 5th power function.

Table 10.2.5 Optimal flow/capacity for different forms of the objective function

	Optimal flow/capacity ratio in p.c.u. per hour per lane for a road in urban area of the Randstad	
Dimension interval	$F = x \left\{ D1 \left(\frac{x}{c}\right)^{7} + D3 \left(\frac{x}{c}\right)^{3} + D2 \right\} + Ac + B$	$F = x \left\{ D1 \left(\frac{x}{c}\right)^{5} + D2 \right\} + Ac + B$
0–4 lanes	1,340	1,390
4–8 lanes	1,220	1,250
8–12 lanes	1,320	1,360

10.2.2 Sensitivity of the Objective Function and the Marginal Objective Function

(a) *For normal intervals*

For 'normal' intervals of the cost function the objective function is given by relationship (10.1.11):

$$F = x \left\{ D2 + D1^{\frac{1}{1+CCRE}} A^{\frac{CCRE}{1+CCRE}} \left(CCRE^{\frac{-CCRE}{1+CCRE}} + CCRE^{\frac{1}{1+CCRE}} \right) \right\} + B$$

$$(10.1.11)$$

The marginal objective function follows immediately from this:

$$\frac{dF}{dx} = D2 + D1^{\frac{1}{1+CCRE}} A^{\frac{CCRE}{1+CCRE}} \left(CCRE^{\frac{-CCRE}{1+CCRE}} + CCRE^{\frac{1}{1+CCRE}} \right) \quad (10.2.4)$$

In the stepwise assignment according to the least marginal objective function, the routes with the lowest value for the marginal objective function are used and constructed. The marginal objective function consists of two parts: $D2$, the fixed part of the users' costs dependent only on the distance and a complicated part, in which all factors of importance in defining the optimal flow/capacity ratio play a role. Both parts are of about the same magnitude. Increasing the investment costs, the value of travel time and/or the discount rate augments the second part, in which the marginal investment costs weigh most heavily. The relative weights of the different components of $D2$ are shown in Table 9.8.2. The influence of $D1$ and A on the second part of the objective function appears from relationship (10.2.4); the precise influences depend on the values for the remaining coefficients.

The influence of CCRE on the marginal objective function is rather complicated. When CCRE is increased the terms

$$D1^{\frac{1}{1+CCRE}} \quad \text{and} \quad \left(CCRE^{\frac{-CCRE}{1+CCRE}} + CCRE^{\frac{1}{1+CCRE}} \right)$$

decrease, but the term $A^{CCRE/(1+CCRE)}$ increases. Differences in the investment costs weight, therefore, more heavily in the second term with increasing CCRE, but on the other hand, the second term weighs more lightly in the total marginal objective function. Sensitivity computations show that the marginal objective function is somewhat less sensitive for differences in the investment costs at increasing values for CCRE. However, this sensitivity of

Table 10.2.6   Sensitivity of the marginal objective function for different investment costs at different values of CCRE

Road type	Marginal objective function (index)	
	CCRE = 5	CCRE = 7
Road in urban area of the Randstad:		
0–4 lanes	121	116
4–8 lanes	100	100
8–12 lanes	115	113
Road in rural area of the remainder of the Netherlands:		
0–4 lanes	100	100
4–8 lanes	95	97
8–12 lanes	115	113
Circular road 0–12 lanes	153	142
Existing road	71	79

the sensitivity is rather small. Table 10.2.6. gives some values for the total marginal objective function for two values of CCRE.

(b) *For other intervals*

In the 'normal' intervals of the cost functions, the coefficient B for the fixed investment costs does not influence the marginal objective function. However, when a transition takes place from a certain interval of the investment function to the next one, the transition from $B1$ to $B2$ or from $B2$ to $B3$ has a considerable influence. Of course, this holds especially when the marginal objective function is given as a difference quotient instead of a differential quotient.

If the dimension constraint becomes active, the objective function is given by relationship (10.1.29):

$$F = x\left\{D2 + D1\left(\frac{x}{c^{\max}}\right)^{CCRE}\right\} + Ac^{\max} + B \qquad (10.1.29)$$

The marginal objective function follows immediately from this:

$$\frac{dF}{dx} = D2 + \frac{(CCRE + 1)D1}{c^{\max\,CCRE}}x^{CCRE} \qquad (10.2.5)$$

It is obvious that the objective function and the marginal objective function increase very rapidly at increasing flow and that the higher CCRE is, the more rapid this increase. This is illustrated in Table 10.2.7, where the

Table 10.2.7 Marginal objective function on a dimension-constrained road ($c^{\max} = 4,000$) for two values of CCRE

Flow	Marginal objective function (index)	
	CCRE = 5	CCRE = 7
2,000	100	100
2,250	100	100
2,500	100	100
2,750	124	140
3,000	160	210
3,250	208	325
3,500	275	507
3,750	363	787
4,000	478	1,200
4,250	625	1,800
4,500	810	2,670
4,750	1,050	3,870
5,000	1,340	5,520

marginal objective function is given for a road constrained to two lanes per direction for the values 5 and 7 for the exponent CCRE, while $D1$ and $D2$ are adjusted to CCRE in the correct way.

(c) *Sensitivity of the network optimization*

Finally, we will say something about the sensitivity of the network optimization, as far as it can be deducted from the sensitivity of the relationship between the number of lanes and the traffic flow and from the sensitivity of the marginal objective function.

The relative weight of the first term of the marginal objective function ($D2$) with respect to the second one in relationship 10.2.4 is especially important. If the first term is relatively high, a strong tendency exists to use and construct the routes with the shortest distances. This can often lead to constructing new roads, which in turn leads to high investment costs, as well as to the overloading of dimension-constrained but well-situated roads. This is because the 'congestion-sensitive' part of the users' costs does not outweigh the fixed part which depends only on the distance.

The total amount of investments is determined to a large extent by the optimal relationship between the number of lanes and the traffic flow. The sensitivity of the total amount of investments consequently will be of the same order of magnitude as the sensitivity between the number of lanes and the traffic flow as discussed in Section 10.2.1 (Table 10.2.1 and Table 10.2.3). Moreover it can be expected that the sensitivity of the total investments is somewhat smaller than that discussed in Section 10.2.1 for situations with much overcapacity (large value for c^{min}).

On the other hand, in situations without overcapacity the sensitivity of the amount of investments will be somewhat greater since for the necessary extra capacity, investments must be made. The characteristic of investments is that integral expansions for marginal dimensions are involved. Of course this effect is strongest if the network has to be 'very slightly' expanded. In addition, the non-linear relationship between investments and dimensions plays a role.

10.3 THE OPTIMAL RELATIONSHIP BETWEEN THE TRAFFIC FLOW AND THE NUMBER OF LANES

When the optimal relationship between the traffic flow and the number of lanes is defined according to relationship (10.1.10) with the values for the coefficients as determined in the preceding chapter, an optimal flow of 1,200 to 1,500 p.c.u. per lane for the average evening peak hour is found for the year 1980; this resulting optimal peak hour flow is 1,175 to 1,475 p.c.u. and 1,150 to 1,450 p.c.u. per lane respectively for the years 1990 and 2000. As is known, the variation in the optimal flow/capacity ratio is caused by the variation in the costs for investments, while the differences per year are caused by the increase in real income.

In Section 10.2 the sensitivity of the result for the different parameters in the objective function has been discussed. The optimal flow/capacity ratio turns out to be not too sensitive for the absolute level of the different cost elements, but fairly sensitive for the assumptions about the form of the objective function (the exponent CCRE) and about the distributions over time of the traffic. In Section 9.2 up to and including Section 9.4, it became clear that the very form of the functions is quite uncertain. Because of these reasons it is important to compare the results mentioned above with figures and standards applied elsewhere.

The necessary or desired number of lanes of a road is usually based on the traffic flow to be expected on the decisive peak hour of the design year. Usually the design year is a year about fifteen years after the opening of the related road and, for the flow on the decisive peak hour, the thirtieth highest hourly flow is usually taken. For highways in the Netherlands in 1970 the thirtieth highest hourly flow generally equals about 12 per cent of the average daily flow, while the flow on the average peak hour equals generally about 10 per cent of the average daily flow.

In the literature one can hardly find any 'standards' for the number of lanes for a certain traffic flow. Winch (1963) gives an extensive list of all cost factors which have to be considered and says that that dimension has to be chosen which minimizes the sum of these costs. However he does not mention any figures. The more technically oriented works of the American Association of State Highways Officials (1965a, 1965b) give only indications such as: 'If this level of service is desirable then such and such traffic flow per lane is implied.' The answer to the question of which level of service is desired is left to the designer or the decision-making authority. Generally level of service B or C for the thirtieth highest hourly flow in the design year seems to be considered desirable for rural highways. That means a flow of 600 to 1,250 p.c.u. per lane for the average evening peak hour. It may be doubted however whether the Dutch highways must be considered as rural highways. For expressways in the Chicago Area level of service C is considered desirable and level of service D permissible for the thirtieth highest hourly flow in the design year (Highway Design Standards Sub-Committee, 1969). That means respectively 850 to 1,250 and 1,250 to 1,500 p.c.u. per lane for the average evening peak hour.

The Rijkswaterstaat in the Netherlands applies the 'standard' that the traffic flow on the decisive peak hour in the design year does not exceed the 1,500 p.c.u. per lane. Only in places where extremely high investment costs are necessary is a traffic flow of 1,800 p.c.u. per lane for the decisive hour in the design year considered permissible. For the average evening peak hour, on which the flow/capacity ratios are based in the Dutch Integral

Transportation Study, this means a traffic flow of 1,250 and 1,500 p.c.u. per lane respectively. This means that a new road is dimensioned in such a way that the expected traffic flow on the average evening peak hour of the design year does not exceed the 1250 p.c.u. per lane. This number is based on the requirement that the traffic operates in a regular way in the majority of evening peak hours. The regular traffic operation is also important with respect to safety on the road. On the other hand a flow of 1,250 p.c.u. per lane for the average evening peak hour implies that on many days a larger, and even much larger, traffic flow will occur, while the flow during short intervals will be much larger too.

With respect to the standards applied elsewhere, it must still be noted that there the expected traffic flow in the design year is always used, while in the Dutch Integral Transportation Study the traffic flow at the moment itself is considered.

Finally, we will mention that in the Netherlands in 1970 the traffic flows on the average evening peak hour hardly anywhere exceeded the 1,100 p.c.u. per lane (Rijkswaterstaat, 1971). This held even for the more heavily loaded direction of the most heavily loaded highways. A few exceptions exist, such as rijksweg (national highway) 16 at Dordrecht (1,780 p.c.u. per lane), rijksweg 4 at Nieuw-Vennep near Amsterdam (1,500 p.c.u. per lane) and rijksweg 8 at the Coentunnel near Amsterdam, too (1,500 p.c.u. per lane). So generally the flow on the average evening peak hour in 1970 lies below our 'optimum'. A flow/capacity ratio above the general 'optimal' one may also occur in the optimized network; at those places where the dimension equals the maximal dimension. On the other hand our 'optimal' flow/capacity ratio is based on a broken number of lanes. Transition to an integer number of lanes may imply a decrease of the flow/capacity ratio.

From the above it can be seen that there is a significant difference between the results of our computation and the standards and figures based on the experience of others. However this difference is not very large. Two possible causes may exist for this discrepancy:

(a) others have not defined the (optimal) relationship between traffic flow and number of lanes on the basis of a minimization of the social costs;
(b) in this study the costs applied are incompletely or incorrectly defined.

With respect to point (a), others not applying a minimization of the social costs, it can indeed be stated that only very rarely does a formal costs minimization take place (it has been done by Winch, 1963, for instance, but he did not come to concrete results). However, the standards used by others *are* a result of the weighing of a reasonable level of service on the one hand and construction and other costs on the other.

With respect to point (b) we may state that the composition of the cost factors applied seems fairly complete (see Section 9.2 up to and including

Section 9.6). Only the environmental costs have not been included in the objective function, but it seems hardly probable that these have a great impact on the optimal flow/capacity ratio. Also, for the amounts of money put down for the different cost factors, reasonable estimates seems to have been made. However, it is extremely difficult to give a very close estimate for all these factors, again because many assumptions must be made which are not always verifiable. However, from the sensitivity analyses of Section 10.2 it has become clear that the result is generally not too sensitive for the absolute level of the different cost factors. Only the distribution of the traffic over time has an important impact.

With respect to the functional relationships giving the relations between the level of the different costs and the traffic flow, we know it to be highly probable that these are not always correct. Not infrequently the shape of the functions has been chosen so that the computational work is made easy. (Of course this is something always done in a study in which the problem is whether further detailing and an attendant increase of costs outweighs the value of the information obtained.) Here too, in some places, the sensitivity of the result for the supposed shape of the function has been studied. It turns out that fluctuations in the results are possible having the same, order of magnitude as, or even greater than, the difference between the result obtained in this study and the empirically determined standards and numbers used by others.

Finally, we must point out in addition that, for the users' costs, only the costs made on workdays have been included in the objective function while the costs for investments and maintenance have been taken in total. If the users' costs for the weekend were also to be included in the objective function, the resulting optimal flow/capacity ratio would be lower (for $D1$ would become larger).

Considering the above, it seems sensible to adjust the result originally computed to the standards and numbers based on the experience of the Rijkswaterstaat and others. Therefore, in the Dutch Integral Transportation Study, the average optimal traffic flow has been stated as 1,250 p.c.u. per

Table 10.3.1 Values for $D1$ ultimately used in the computation process (in Dutch guilders of 1970 per kilometre)

Year	$D1$
1980	293·12
1990	336·29
2000	370·25

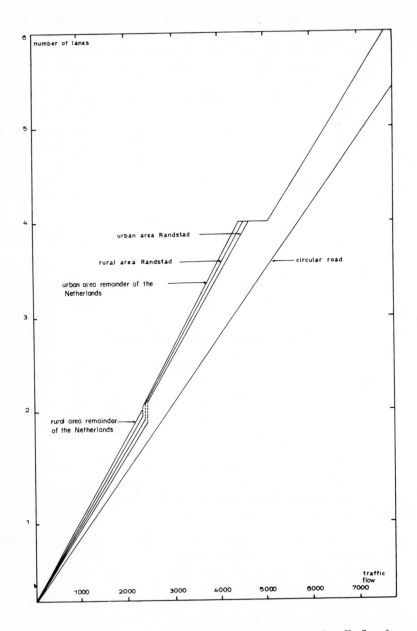

Figure 10.3.1   Optimal relationship between dimension and traffic flow for the five types of road

lane for the average evening peak hour, while the fluctuations in the optimal traffic flow caused by the difference in investment costs have been maintained at between *1,100 and 1,400 p.c.u. per lane for the average evening peak hour*. This means that the originally resulting optimal number of lanes for a given traffic flow is increased by 7·5 per cent. Computationally this is realized in the sub-problems for the network optimization by multiplying $D1$ by $(1·075)^6$. The ultimately resulting values for $D1$ are given in Table 10.3.1.

In Figure 10.3.1 the optimal number of lanes is given as a function of the traffic flow on the average evening peak hour for the five main types of road. Moreover the optimal flow per lane for the average evening peak hour is given in Table 10.3.2 for three types of road and for the special structure with the very highest (marginal) construction costs, namely a bridge in the Delta-area.

Table 10.3.2   Optimal traffic flow (in p.c.u.) per link for the average evening peak hour (for the year 1980)

Type of road ⟍ Dimension interval	Circular road (5 km, normal intersections)	Road in urban area of the Randstad (5 km, normal intersections)	Road in rural area of the remainder of the Netherlands (5 km, normal intersections)	Bridge in the Delta-area (special structure with 2 km normal road)
0–4 lanes	1,430	1,295	1,165	1,690
4–8 lanes	1,430	1,160	1,110	1,375
8–12 lanes	1,430	1,265	1,265	1,430

10.4   FROM CONTINUOUS DIMENSION TO A DISCRETE NUMBER OF LANES

Up to now the dimension of a road has always been considered as a continuous variable. There are reasons for this:

(a) A fractional number of lanes gives more information than an integral number of lanes. A road with 3·1 lanes has to be expanded to a 4-lane road later than a road with 3·7 lanes. This is especially relevant when the optimization is executed for certain dates in order to define the optimal dimension for the whole time period. However, one may object that the extra information can also be obtained from the traffic flows.

(b) Using a continuous dimension, the minimum can be obtained by differentiation (Section 10.1.2.3). Though this is, of course, an advantage, the advantage must not be overvalued. Owing to the clearly convex shape of the total costs for a given traffic flow as a function of the dimension, the optimal dimension can be found easily by a simple search method (Section 10.1.2.2).

(c) For the successful application of the stepwise assignment according to the least marginal objective function, it is necessary that the objective function $F(x)$ be a continuous function of the traffic flow x. If we change the dimension from c_n to c_{n+1} for that x for which $F(x, c_n) = F(x, c_{n+1})$, the function $F(x)$ is indeed, a continuous function of x, also if c_n is a discontinuous variable except for the point $x = 0$, where the function has become discontinuous now:

$$F(0) = 0$$

$$\lim_{\Delta x \downarrow 0} F(x) = i(c_1)$$

Further the objective function must preferably be convex. But using the method above every jump in the dimension would cause a temporary decrease of the derivative of $F(x)$ with respect to x, that is to say a concavity:

$$\lim_{\Delta x \uparrow 0} \frac{F(x + \Delta x) - F(x)}{\Delta x} = D2 + \frac{D1(CCRE + 1)x^{CCRE}}{c_n^{CCRE}}$$

$$> D2 + \frac{D1(CCRE + 1)x^{CCRE}}{c_{n+1}^{CCRE}} = \lim_{\Delta x \downarrow 0} \frac{F(x + \Delta x) - F(x)}{\Delta x}$$

$$(c_{n+1} > c_n)$$

(d) The real dimension, which is of course an integral number of lanes or the exact date for expansion of the road, can also be defined on the basis of considerations other than the minimization of the social costs on that specific road. For the road forms a part of a whole network.

For a first determination of the amount of investment necessary for a certain year it may be useful to give the road dimensions in an integral

Table 10.4.1   Optimal flow intervals for the average evening peak hour on an integral number of lanes, for a road with an optimal traffic flow of 1,250 p.c.u. per lane for the average evening peak hour when the number of lanes is considered as a continuous dimension

Flow interval in p.c.u.	Number of lanes
0	0
0–1,400	1
1,400–2,980	2
2,980–4,270	3
4,270–5,520	4
5,520–6,800	5
>6.800	6

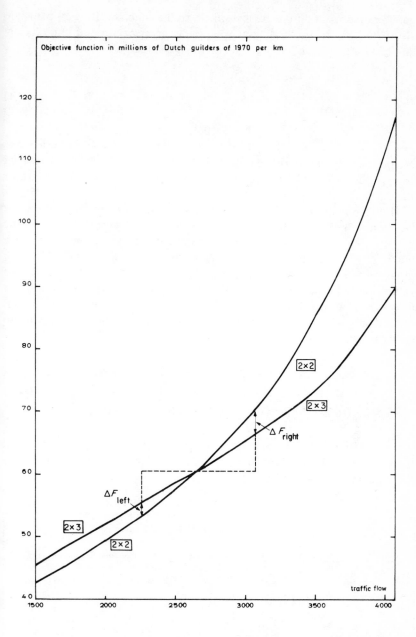

Figure 10.4.1 Objective function (road in an urban area of the Randstad, length 5 kilometres) for a 2 × 2-lane road and 2 × 3-lane road

number of lanes and to relate investment to that number. It seems plausible also to define the traffic flow at which it is necessary to add one lane by minimization of the objective function. So a road has n lanes if:

$$i(c_n) + T(x, c_n) \leqslant i(c_{n+1}) + T(x, c_{n+1})$$

and:
$$(10.4.1)$$

$$i(c_n) + T(x, c_n) \leqslant i(c_{n-1}) + T(x, c_{n-1})$$

For a road with an optimal traffic flow of 1,250 p.c.u. per lane for the average evening peak hour the critical flows have been defined at which the road must be expanded with this relationship (10.4.1). The results are given in Table 10.4.1.

The above approximation does indeed give the minimum of the objective function if the flow is uniquely defined. However, when x is a stochastic variable, the dimensions above do not give the lowest value for the objective function. For too low a dimension is much more serious than too high a dimension. The reader will see how in Figure 10.4.1 the difference in the values for the objective function on the right of the intersection is much larger than on the left. This shows that the road must be expanded earlier.

Assuming a distribution for x with a density function $f(x)$ relationship (10.4.1) changes into:

$$i(c_n) + \int_0^\infty f(x)\, dT(x, c_n) \leqslant i(c_{n+1}) + \int_0^\infty f(x)\, dT(x, c_{n+1})$$
$$(10.4.2)$$
$$i(c_n) + \int_0^\infty f(x)\, dT(x, c_n) \leqslant i(c_{n-1}) + \int_0^\infty f(x)\, dT(x, c_{n-1})$$

The dimension intervals have been computed, assuming for $f(x)$ a normal distribution with a standard deviation of 10 per cent and a standard deviation of 20 per cent (Table 10.4.2).

Table 10.4.2   'Rounding' on an integral number of lanes

Discrete number of lanes	Dimension intervals in fractional number of lanes		
	x deterministic	x normal distributed	
		$\sigma = 10$ per cent	$\sigma = 20$ per cent
0	0	0	0
1	0–1·12	0–1·10	0–1·05
2	1·12–2·38	1·10–2·35	1·05–2·24
3	2·38–3·41	2·45–3·36	2·24–3·21
4	3·41–4·43	3·36–4·36	3·21–4·17
5	4·43–5·44	4·36–5·35	4·17–5·12

10.5 OPTIMIZATION OVER TIME

In the approach applied to the Dutch Integral Transportation Study three networks were defined for three different years (see Chapter 8). In this case three static problems had to be solved. In Section 6.1 the dynamic network optimization problem has been discussed. In Section 6.1.3 the dynamic road optimization problem has been stated as a subproblem of the network optimization. Here we will say something more about this dynamic road optimization problem.

In the simplest case there are two possible dimension levels: c_n and c_{n+1}, with related investment costs $i(c_n)$ and $i(c_{n+1})$ and also given investments costs Δi for the expansion from c_n to c_{n+1}. The objective function is considered for a period from the starting date t_0 till the final data t_e, while the expansion takes place on date t^*. The objective function consists of the total costs, all discounted to the present value:

$$
F = i(c_n) + i\,e^{-\pi t^*} + \int_{t_0}^{t^*} T(x(t), c_n)\,e^{-\pi t}\,dt
$$

$$
+ \int_{t^*}^{t_e} T(x(t), c_{n+1})\,e^{-\pi t}\,dt
$$

(10.5.1)

The question is to define the optimal date t^* for the expansion:

$$
\min_{t^*} F(t^*)
$$

If the flow $x(t)$ is a continuously increasing function of t, $F(t^*)$ is a convex function, so the minimum can be found by setting the differential quotient of F with respect to t^* equal to zero, unless the minimum lies on the boundary as when $t^* = t_0$ or $t^* = t_e$. But these possibilities can be studied easily. Differentiation of relationship (10.5.1) and putting it equal to zero gives:

$$
-\pi\Delta i\,e^{-\pi t^*} + T(x(t^*), c_n)\,e^{-\pi t^*} - T(x(t^*), c_{n+1})\,e^{-\pi t^*} = 0
$$

or:

$$
\pi\Delta i = T(x(t^*), c_n) - T(x(t^*), c_{n+1})
$$

So the loss in users' costs must equal the product of the investments for the expansion and the discount rate to make expansion profitable. This result is well known in transportation research. The traffic flow at which the road must be expanded fully corresponds, of course, with the flows given in Section 10.4. From the flow the date t^* can be easily computed.

The problem becomes more complicated when there are more possible dimension levels, more possible dates for expansion and the traffic flow is not a continuously increasing function of time. In that case an optimal series of investments must be defined. The problem can be stated as follows:

$$\min_{c_1, \ldots, c_n} \left\{ \sum_{t=1}^{n} T(x_t, c_t) e^{-\pi t} + \sum_{t=1}^{n} i(c_t - c_{t-1}) e^{-\pi t} \right\}$$

When the number of variables is not too large this problem can be solved easily by dynamic programming (see Section 6.1.2). For this the following recurrence relations can be formulated:

$$F_1(c_1) = T(x_1, c_1) e^{-\pi} + i(c_1).$$

$$F_t(c_t) = \min_{c_{t-1}} \{F_{t-1}(c_{t-1}) + T(x_t, c_t) e^{-\pi t} + i(c_t - c_{t-1}) e^{-\pi t}\}$$

$$\text{for } t = 2, \ldots, n$$

$$F(c_n) = \min_{c_n} F_n(c_n)$$

REFERENCES

American Association of State Highway Officials (1965a). *Traffic Engineering Handbook*, 3rd edition, A.A.S.H.O., Washington, D.C.
American Association of State Highway Officials (1965b). *Policy on Geometric Design of Rural Highways*, A.A.S.H.O., Washington, D.C.
Hardy, G. H., Littlewood, J. E., and Polya, G. (1934). *Inequalities*, Cambridge.
Hendriks, A. J., and Vloemans, A. W. H. G. (1971). *Enige Sociaal-economische Aspecten van de Maatschappelijke Ontwikkeling in de Komende Decennia*, Deelrapport 11a, Integrale Verkeers- en Vervoerstudie door het Nederlands Economisch Instituut, Rotterdam.
Highway Design Standards Sub-Committee (1969). *Design Criteria for Expressways in the Chicago Metropolitan Area*, Chicago.
Highway Research Board (1966). *Highway Capacity Manual, 1965*, Washington D.C.
Rijkswaterstaat, Dienst Verkeerskunde (1971). *Verkeerstellingen in 1970*, Rijkswaterstaat, 's-Gravenhage.
Wilde, D. J. (1964). *Optimum Seeking Methods*, Prentice-Hall.

11

Application of the Stepwise Assignment According to the Least Marginal Objective Function to Get the Optimal Road Network for a Given Trip-matrix

11.1 BASIC DATA

11.1.1   The Trip-matrix

The matrices of the numbers of trips by car are determined in the transport forecast of the Dutch Integral Transportation Study (see annex III and annex IV of the final report of this study). The number and the direction of the trips depend on a large number of variables, of which the size and geographical location of population, working population and employment (see annex II of the final report of the Dutch Integral Transportation Study) and income and car-ownership (see annex I of the report mentioned) are the most important ones. Matrices for the trips between 351 zones, therefore having 351 rows and columns, are employed. The users' costs for the traffic on workdays are defined. This implies that a trip-matrix for the workday must be used. In Sections 9.2, 9.3 and 9.4 it has been said already that the total users' costs are computed from the flows in the average evening peak hour. For this it is necessary to know the distribution over time of the traffic flows for every road.

For practical reasons the same coefficients for the users' costs are used for every road. For this it is necessary to assume that the distribution over time is equal for all roads. This last assumption can be based on the following two assumptions:

(a) the distribution over time is equal for all relations;
(b) the route choice (according to a normative or descriptive assignment) is equal for every point of time.

Neither assumption is fully correct and so it is important to test the sensitivity of the results for these assumptions. Neither assumption would be necessary if for every hour (minute) an estimate for the traffic were made and for these trip-matrices an optimization system were applied as discussed in Section 6.1.4. Obviously this is only a theoretical possibility. The trip-matrix for the average evening peak hour is now used to estimate the total users' costs. The evening peak hour has been chosen for two reasons:

(a) a large part of the total users' costs, and especially of the marginal users' costs, is incurred in this hour;

(b) the traffic flows in the second busiest hour, the morning peak hour, on the whole greatly resemble the flows in the evening peak hour (except in their direction).

The first point is illustrated by the value of the total travel time losses per year if, in the average evening peak hour, one p.c.u. incurs a travel time loss of one hour. This value (D1TIME/CCR as discussed in Section 9.2.3) works out at about 3,200 Dutch guilders for 1980, apart from the fluctuations within the hour, the week and the month and fluctuations determined geographically. If the travel time losses were to occur only in the evening peak hours, this value would have been $250 \times 4 \cdot 50 = 1,125$ Dutch Guilders, again apart from the fluctuations mentioned (with 4·50 Dutch Guilders for the average value of one hour travel time loss for one p.c.u. in 1980). This means that about 35 per cent of the travel time losses incur during the evening peak hour (N.B. *not* 35 per cent of the *total* travel time, this value being, of course, less interesting). So to base our figures on the evening peak hour seems all right, even if there were to be large differences between the distributions of the traffic over the day.

To get an idea of the effect of this choice of the evening peak hour and the related assumptions, the optimization has also been carried out for a trip-matrix of the average daily traffic, divided by a factor such that the total number of driven kilometres equalled the total number of driven kilometres in the average evening peak hour. For the rest, the same coefficients in the cost functions and the same parameters for the computation process have been applied.

This computation resulted in a value of the objective function about 6 per cent lower than for the optimization based on the peak hour. This lower value is caused by a more uniform distribution of the traffic over the country during the day whereas, in the evening peak hour, high traffic concentrations are present around the big cities. This last fact is the cause of high congestion costs and high investments. The amount of investments computed for the 'daily trip-matrix' is about 4 per cent lower than that for the evening peak hour trip-matrix (i.e., 4 per cent when the number of lanes is not necessarily equal in both directions).

After this consideration of the macroresults, an examination of the networks is especially important. Then we see the places where the differences appear and whether there are more differences than are shown by the macrociphers. Of course the flows in both directions equal each other almost completely for the daily matrix whereas this is not the case for the evening peak hour. However, for the rest there are no large differences in the flow pattern for the two matrices. The most important differences appear around the large towns, while for the remainder of the Netherlands the patterns are almost identical.

Because of the above it seems justified to use the trip-matrix for the car traffic in the evening peak hour to make an estimate of the total users' costs of the traffic on workdays.

The trip-matrix used* is the summation of a forecasted trip-matrix of the national traffic by car of persons and a forecasted matrix of national and international goods transport converted first to loaded and unloaded lorries and next to passenger car units. Next a small universal increase in the matrix has been applied to take into account the person traffic across national frontiers (including the traffic of travellers to and from the airports), the public transport using the road and that part of the traffic inside the transport zones that uses the highway network. Of course this last increase in the matrix is a poor approximation of those parts of the traffic not included in the forecast. But fortunately only a small percentage of the total traffic is concerned.

In the Dutch Integral Transportation Study only the interurban infra-structure is considered. So only the traffic to and from the towns is relevant. For forecasting however it is necessary to divide the large towns into a number of transport zones. This is important too for the road network, because it may make some differences if a trip is destinated in the north or the south of a large town. The traffic between the different zones of the same city is computed in the forecasting as well. These trips are not directly relevant for the optimization of the interurban road networks (at least under certain assumptions for the urban and interurban road networks, see further Section 11.1.2) and can even be disruptive. So the traffic for the urban relations have been put equal to zero for the optimization process.

11.1.2 The Network

As is known, the position of all possible roads and their maximum dimensions form data for the optimization, while in the optimization process itself a choice is made from these given possibilities and the dimensions of the different roads are defined. To define the given maximum structure, the structure scheme of the highway network for the Netherlands given by the Ministry of Transport has been taken into account (*Struktuurschema*

---

* The trip-matrix used and described in this chapter formed a preliminary forecast of the Dutch Integral Transportation Study for the year 2000. This matrix was later improved. We still use this preliminary matrix in this book because more test material is available for this matrix. For a description of the results of the final trip-matrices see annex V of the final report of the Dutch Integral Transportation Study.

Nederlandse Hoofdwegennet, 1966, to be found in the *Tweede nota over de ruimtelijke ordening in Nederland* (1966, page 138)). The important existing secondary roads have also been taken into the network in abstracted form. Moreover roads have been added where there were gaps before, where special developments were possible and where important plans were present on a lower governmental level. Furthermore the maximum structure has been constructed in such a way that every nucleus is connected to the existing network and that two adjacent nuclei are connected to each other in a relatively easy way.

The resulting network is described in a number of nodes and links. These nodes and links can be distinguished as:

(a) *Centroids:* nodes where the traffic is assumed to have its origins and destinations. Every transport zone contains one centroid.

(b) *Ordinary or intermediate nodes:* all nodes (points in the network where two or more links come together) which are not centroids. These ordinary nodes may represent an intersection or a special structure; they may also only be taken into the network to connect a centroid connector (see below) with the network or to indicate the coordinates of a point of the road, so that the length of a curved road can be computed more accurately and the road can be plotted better by the computer.

(c) *Roads:* links between two ordinary nodes. These roads represent in abstracted form the real (possible) road network. Every road bears some characteristics. The most important one is whether it is included in the optimization process or not.

(d) *Centroid connectors:* links which connect a centroid with an ordinary node or another centroid. So the centroid connectors serve only to connect the centroids of the transport zones with the road network and do not represent real roads.

The numbers of nodes and links of the network used are given in Table 11.1.1.

Table 11.1.1   Numbers of nodes and links of the network

Number of ordinary nodes	1,549
Number of centroids	351
Number of roads[a]	2,433
Number of centroid connectors[a]	624

[a] Because both directions of the links are treated separately during the computations the total number of links is 6114.

For all roads a lower and an upper bound for the dimension is stated. The lower bound equals the dimension assumed for the year 1975; expansion beyond the upper bound is not assumed to be possible or sensible (see further Section 10.1.4).

Special measures must be taken for the roads not included in the optimization process. The roads inside towns and the roads abroad belong to this category. These roads must be treated in such a way that the computations for the other roads are not disturbed. The following is assumed for the roads inside towns:

(a) traffic not originating or destinating in the related town does not use the road network of that town, unless there are no other reasonable possibilities;
(b) traffic originating or destinating in the related town, uses the road network inside and around the town in a realistic way.

These assumptions must be made operational for the computation process. For the stepwise assignment according to the least marginal objective function this means defining a value for the criterion $F_{ij}(x_{ij} + \Delta x) - F_{ij}(x_{ij})$ for the related roads. For the descriptive assignment (according to the shortest travel time in which the speed depends on the flow/capacity ratio: capacity restraint) the speed on the related roads must be defined. In order to define the value for the criterion $F_{ij}(x_{ij} + \Delta x) - F_{ij}(x_{ij})$ the value for this criterion for the surrounding roads is considered, especially the circular roads in relation with the roads inside towns. In Table 11.1.2 the value of the

Table 11.1.2 Marginal objective function per kilometre for a new circular road with maximally four lanes per direction as a function of the traffic flow for two values of Δx

Traffic flow x in p.c.u.	Flow per lane	$F(x + 1967) - F(x)$ thousands of Dutch guilders of 1970	$F(x + 1311) - F(x)$ thousands of Dutch guilders of 1970
2,622	1,311	14,557	9,705
3,933	1,311	14,950	9,705
5,244	1,311	31,058	14,370
6,555	1,639	99,820	48,897
7,866	1,993	289,034	142,570

marginal objective function for a new circular road with a maximum of four lanes for one direction is given (see Section 11.1.3 for the coefficients used. These coefficients do not equal the values given in Chapter 9). On the basis of the values of Table 11.2.1 the assumptions about the use of the roads inside towns have been assigned a value of 100,000 per kilometre for ΔF (in thousands of Dutch guilders of 1970).

At the descriptive assignment, a constant speed of 30 km/h has been assumed for all roads inside towns. For the speed/flow relationship used at the descriptive assignment, this speed is reached when traffic flow is about

1,350 p.c.u. per hour per lane. This constant speed for all roads is a rather rough assumption. A system with differentiated speeds would have been better. The speeds could have been chosen in such a way that the travel time is equal for every closed ring around the centre, while the speeds for the radial roads are adjusted to that.

For the roads abroad the same values for the criterion ΔF and for the speed in the descriptive assignment have been chosen as for the roads inside towns. This choice is of course questionable.

The centroid connectors must be treated in a different way from the roads mentioned above. The following conditions can be stated:

(a)   only traffic originating or destinating in the related centroid uses the centroid connector;
(b)   the length of the centroid connectors does not influence the routes to be chosen.

To satisfy these conditions a value has been chosen for the criterion ΔF which is very high and which is also valid for the centroid connector as a whole: that is 5,000,000 for the whole centroid connector. For the travel time in the descriptive assignment, a high value is also chosen, namely 200 minutes, again for the centroid connector as a whole. Note that the values selected cannot be too high, otherwise during the computation process too much accuracy is lost for the other roads.

11.1.3   Values Used for the Coefficients of the Objective Function

For the construction and maintenance costs, the cost functions with coefficients of Section 9.5 have been used. In the test computation being discussed here, for the total users' costs a function has been used with the same mathematical shape as discussed in Chapter 9, but with other values for the coefficients, namely:

$$T = x \left\{ D2 + D1 \left(\frac{x}{c} \right)^{CCRE} \right\}$$

with:

$D1 = 840$ (Dutch guilders of 1970 per kilometre)

$D2 = 280$ (Dutch guilders of 1970 per kilometre)

$CCRE = 7$

The reason for describing the test computation instead of the final one is that more material is available for this test computation (see Chapter 4 of annex V of the final report of the Dutch Integral Transportation Study for a discussion of the final computations). Later on in the Dutch Integral

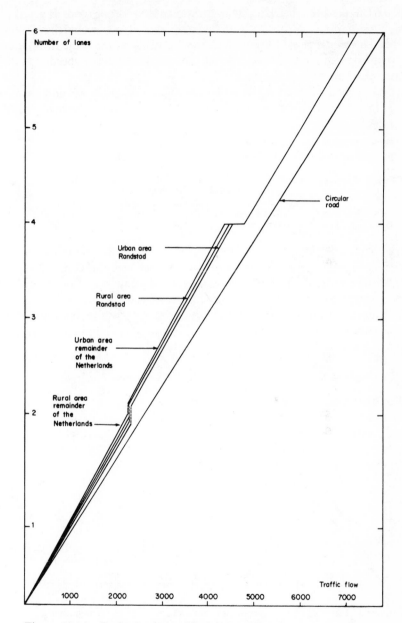

Figure 11.1.1 Optimal relationship between dimension and traffic flow for the five types of road in the test computation

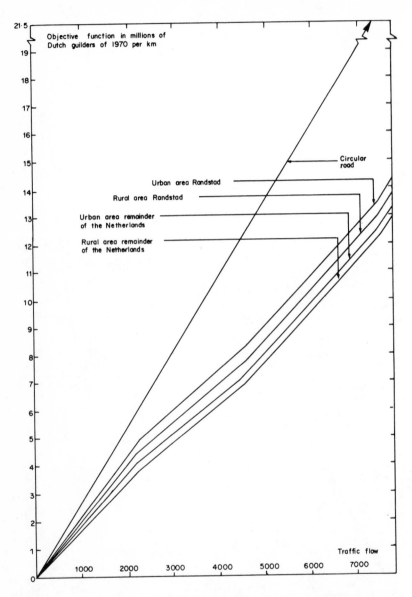

Figure 11.1.2   Objective function for the five main types of road in the test
computation

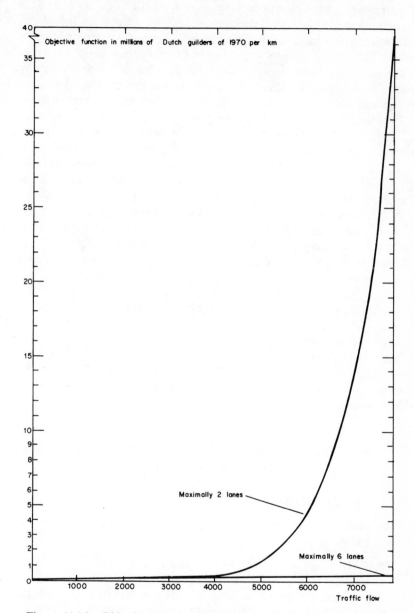

Figure 11.1.3   Objective function for dimension-constrained road in the
test computation

Transportation Study it was decided to change to the fifth power for CCRE, because that gave a better description for all elements of the objective function together (the seventh power giving a better description of the travel times). The optimal relationship between the traffic flow and the number of lanes almost coincides with the relationship mentioned in Section 10.3 (see Figure 11.1.1), as the relationship between the objective function and the marginal objective function and the traffic flow agree fairly with the relationships described in Section 9.8 (see Figure 11.1.2 and 11.1.3). Only near the dimension constraint does the seventh power function increase more rapidly with increasing flow than the fifth power function given in Chapter 9.

In Figure 11.1.2 and 11.1.3 the required convexity for the objective function is shown clearly for flow values above 2,500 p.c.u. in the average peak hour. Especially when approaching and/or 'exceeding' the dimension constraint, the function is clearly convex. For roads with an existing dimension of 2×2 or more lanes, the function is convex for the whole area. Unfortunately the function is not convex for a traffic flow lower than about 2,500 p.c.u. in the average peak hour. Figure 11.1.2 shows, however, that the non-convexity which appears is not very dramatic.

In the above discussion, the values of the coefficients applied to the optimization process have been demonstrated. Once this process has been carried out, the road network and the flow pattern on it are known. Of course, it is then possible to recompute the value of the objective function for which the requirements of continuity and convexity are no longer necessary. So for the 'final' computation of the amount of investments, slightly different coefficients for the investment costs of roads with two or less lanes for one direction than those given in Section 9.5.2 have been used.

11.2 THE COMPUTATION PROCESS USED AND THE VALUES OF ITS PARAMETERS

11.2.1 The Computation Process Used

The final computation process used consists of three parts (subroutines for the computer program):

(a) SALMOF: *stepwise assignment according to the least marginal objective function* as discussed in Chapter 5;

(b) DESCASS: *descriptive assignment* of the trip-matrix to the network according to the routes with minimum value for the travel time in which the speed is a function of the flow/capacity ratio (capacity restraint as discussed in Section 2.1.4.2;

(c) OPTADJ: *optimal adjustment* of the dimensions of the links to the traffic flows.

For the subroutines SALMOF and OPTADJ use is made of the shortest path algorithm of Moore/Fore/Bellman as discussed in Section 7.2.1.

The computer program has been coded in FORTRAN IV and has run on an IBM 360/65 computer. For the network mentioned in Section 11.1.2 a computer storage of about 500 kilobytes is necessary. The subroutines SALMOF and DESCASS each use 10 to 15 minutes computation time per step; for the subroutine OPTADJ only a few minutes computation time are necessary.

Three computation processes have been studied, all consisting of the subroutines mentioned above:

(a) DESCASS to the maximally dimensioned network, followed by OPTADJ; this is the first step of the process of the continuous optimal adjustment as discussed in Section 4.4.2.1;
(b) SALMOF; the simple application of the stepwise assignment according to the least marginal objective function;
(c) SALMOF followed by a descriptive assignment by DESCASS to the network determined by SALMOF and then again followed by OPTADJ (see Figure 11.2.1 below).

The most important macroresults are given in the indices in Table 11.2.1. In this section we will place the results of the simple application of SALMOF against the two other computation processes, so that we will have compared the results of the first with the second computation process and those of the

Table 11.2.1 Macroresults for three computation processes (in indices)

Function	DESCASS/ OPTADJ	SALMOF	SALMOF/ DESCASS/ OPTADJ
Objective function	114·8	144·4	100·0
Investments in two-lane roads	11·9	3·5	6·3
Investments in multilane roads	46·3	57·6	42·5
Investments in structures	54·1	71·2	51·2
Total investments	112·3	132·3	100·0
Kilometrage on two-lane roads	15·3	9·8	16·9
Kilometrage on multilane roads	84·3	104·3	83·1
Total kilometrage	99·6	114·1	100·0
Average speed on two-lane roads	59·1	27·3	100·0
Average speed on multilane roads	99·8	88·9	100·0

third with the second. A comparison of the results of the first and the third processes will be left to Section 11.4.

It seems at first sight very strange and alarming that the simple application of SALMOF yields the highest value of the objective function, inasmuch as it is precisely SALMOF in which routes are chosen in such a way that the lowest value of the objective function is obtained. Also it seems strange that the investments are highest when they have been taken into account in defining the routes (compare the first with the second column of Table 11.2.1). It is indeed clear that we cannot simply apply the stepwise assignment according to the least marginal objective function for this network optimization problem, at least not with the parameters used.

The high value for the objective function in the simple application of SALMOF is caused by three facts:

(a) Many existing two-lane roads, which are not allowed to be expanded further, are not loaded in SALMOF. This fact is also shown by the smaller number of kilometres driven on two-lane roads in SALMOF. This absence of loading is caused by the high value of the marginal objective function for these roads (see Table 11.2.2).

Table 11.2.2   Value of the marginal objective function for some types of roads in an urban area of the Randstad (in thousands of Dutch guilders of 1970)

	Value of $F(\Delta x) - F(0)$ per kilometre	
Road type	$\Delta x = 1967$ (first step)	$\Delta x = 1311$ (next steps)
Non-existing road, expandable to 6 lanes	8,284	5,522
2-lane road, expandable to 6 lanes	6,539	3,777
2-lane road, not expandable	72,696	5,510

The table shows that, in the first step of SALMOF, the probability that a non-expandable two-lane road is used is very small indeed. In the next steps, such a road is almost as 'expensive' as a new road or as a road where the dimension constraint is approached or where further expansion involves high costs (see also Section 9.8).

Often not using existing two-lane roads causes 'detours' (see the increase of the number of kilometres driven). These 'detours' raise the users' costs. Moreover extra investments in (multilane) roads and structures are caused.

(b) For the road networks inside towns, high values for the loading criterion ('marginal objective function') have been chosen (see Section 11.1.2).

This causes an unusual use of routes in SALMOF with long 'detours' around large towns; this effect is magnified by the crowding on the urban circular roads*

For instance around Amsterdam this situation occurs very clearly. In SALMOF as well as in DESCASS, the circular road around Amsterdam is heavily loaded. In SALMOF there is hardly any traffic through the city, whereas a very heavy flow exists on a half circle about 15 kilometres to the north-east (from Muiderberg over rijksweg 6 and rijksweg 27 through Zuidelijk Flevoland via Marken to Amsterdam). This flow exists for a large part of the traffic from the east (especially the Gooi area) proceeding to the north and west of Amsterdam where there is a much shorter way through the city. In the north of Amsterdam, fewer roads have got the code 'road inside town' than in the south, so that the long detours to avoid the southern part of the city in SALMOF are paying off. Something similar occurs south of Amsterdam. Here traffic is pushed away to roads about 10 kilometres south of the southern circular road (from the south of the city and the circular road, rijksweg 10, to the rijksweg 6 and even rijksweg 80 between Aalsmeer and Mijdrecht). Finally the same occurs in the west. The number of crossings of the Noordzeekanaal west of Amsterdam is about 30 per cent higher in SALMOF than in DESCASS for both directions.

In DESCASS the speed on the roads inside towns almost equals that on the crowded circular roads (see Section 11.1.2). So the road networks inside the towns are then used and the strange detours vanish. These 'detours' again increase the users' costs and often cause high investments. These investments are especially high in this case, because 2×6-lane roads are mostly involved here. Moreover the value of the objective function is increased additionally compared with that of DESCASS, because the costs on the roads inside towns are not included in the objective function.

To what extent the decrease in the value of the objective function in the phase DESCASS/OPTADJ is also justified—in other words whether or not the urban road networks really can handle this more-or-less through-going traffic—is a problem which must be investigated further, together with all other urban transportation problems.

(c) The value of the marginal objective function increases only when approaching the dimension constraint, but it increases very steeply then (see Section 9.8 and Figure 11.1.3). This effect is discussed further in the following subsections. Anyway this implies in some cases that dimension-

---

* In the final computations of the Dutch Integral Transportation Study, a relatively lower value has been taken for the loading criterion for the roads inside towns. The effect mentioned above is reduced; also the value of the objective function for SALMOF is less high compared with the value of the objective function for DESCASS.

constrained (but often existent) roads are not loaded sufficiently and in other cases that roads are overloaded. These overloadings are shown by the low average speeds, especially on two-lane roads.

Because of the three effects mentioned above, a simple application of the stepwise assignment according to the least marginal objective function cannot be used for the problem stated, at least with the optimization parameters used. As has been shown by the results of the three test computations, the objections mentioned above diminish when SALMOF is followed by a descriptive assignment of the trip-matrix to the network and by an optimal adjustment of the dimensions of the roads to the traffic flows on them. Moreover, a descriptive assignment describing reality is used then instead of a normative one, which is (perhaps) another advantage (compare Section 5.5).

The results of the computation process containing a stepwise assignment according to the least marginal objective function are better than the results of the process not containing it, so it indeed seems better to use this optimization procedure. For a deeper analysis of the differences between the results with and without SALMOF, the reader is referred to Section 11.4.

Finally, an iterated application of the process DESCASS and OPTADJ is not thought to be sensible, especially in view of the high computation costs involved.

So in conclusion the third computation process was applied, consisting of a stepwise assignment according to the least marginal objective function followed by a descriptive assignment of the trip-matrix to the network as found before and then completed by an optimal adjustment of the dimensions of the roads to the flows found at the descriptive assignment. A flow chart of this process is shown in Figure 11.2.1.

Figure 11.2.1   Computation process used

11.2.2   Choice of the Number of Steps and Magnitude of the Parts Assigned in SALMOF

In Section 5.7.3 the sensitivity for the different parameters at the stepwise assignment according to the least marginal objective function is investigated for a certain network and trip-matrix. There it was shown that the number of steps must be chosen in accordance with the other parameters and that a small number of steps can be sufficient when the other parameters are properly chosen. For the case discussed in Section 5.7.3, the application of four steps gave relatively very good results. However it was also stated in that section that the right number of steps always depends on the problem stated and the requirements demanded of the accuracy of the solution. So some analysis and experimentation with the problem itself is always necessary. Moreover, the necessary computation time becomes very relevant for networks and trip-matrices of a large size such as are dealt with here. Again, an improvement of the accuracy must always be weighed against the concomitant increase in computation costs.

The objective function used in the Dutch Integral Transportation Study is partly concave (for small values of the dimension) and partly convex. Apart from jumps in the slope, the function is partly linear in x and partly of the (CCRE + 1)th power of x (that means of the sixth power for the final computations and of the eighth power for the computation discussed here). So we encounter a rather complicated form. Generally the use of many small steps at the end of the process seems preferable. On the other hand no step can be too large and this fact tends to make the use of more-or-less equal steps preferable, because the first steps would become too large when only a few steps had been applied.

Table 11.2.3   Effect of the number of steps in SALMOF on the most important macroresults for a test-matrix (in indices)

Function	SALMOF 3 steps	SALMOF 4 steps
Objective function	155·7	100·0
Investments in two-lane roads	11·9	11·0
Investments in multilane roads	37·2	38·3
Investments in structures	48·1	50·7
Total investments	97·2	100·0
Kilometrage on two-lane roads	19·7	19·0
Kilometrage on multilane roads	80·0	81·0
Total kilometrage	99·7	100·0
Average speed on two-lane roads	64·7	100·0
Average speed on multilane roads	36·5	100·0

Table 11.2.4   Effect of the number of steps in SALMOF on the traffic flows on some roads

Road	Traffic flow in p.c.u. in indices on the average evening peak hour	
	SALMOF 3 steps	SALMOF 4 steps
Rijksweg 24; Utrecht–Rotterdam	100	131
Existing road Den Bosch–Eindhoven	100	74
New road Den Bosch–Eindhoven	100	106
Rijksweg 16; near Alphen	100	129
Secondary road 1; near Wassenaar	100	197

Furthermore the results of the total computation process must be considered and not just those of the stepwise assignment according to the least marginal objective function. If a new road has got 'too' small a dimension in SALMOF, this road will get a small flow in DESCASS as well and so too will it have a small dimension in OPTADJ. Thus such an 'error' made in SALMOF is not corrected later, so in any case a good solution in SALMOF is desired on this point.

To get a better idea, the operation of three and of four steps has been investigated for a test trip-matrix.* For the application of three steps, at first a 0·4 part and next twice 0·3 parts of the trip-matrix were assigned with the values respectively for Δx of 1967, 1311 and 1311. For the application of four steps, a quarter of the trip-matrix was always assigned, with the values for Δx of 1967, 1311, 1311 and 1311 respectively.

The most important macroresults of these two experiments are given in Table 11.2.3. The total number of kilometres driven and the amount of investment are a little higher with the application of four steps than with the application of three steps. The difference in the investments is about 3 per cent. Also, the value for the objective function is considerably lowered by the application of four steps. With the larger number of steps and the relative increase of Δx related to this, the response to the dimension constraints is better. The reader is also referred to the slight shifting in kilometrage and in investments from two-lane to multilane roads. In considering the networks and the flow patterns, the differences do not appear to be too large. It was not to be expected from a consideration of the small differences in kilometrage and in investments. However, a few small differences which are important in the next phase (DESCASS/OPTADJ) attract attention.

So some important new highways get more traffic flow in the SALMOF with four steps than with three steps (for instance, the southern highway from Utrecht to Rotterdam (rijksweg 24), the new highway from Den Bosch to

---

* Not the trip-matrix on which the computations discussed in this chapter have been carried out, but nonetheless a matrix with 351 rows and columns.

Eindhoven (rijksweg 2), rijksweg 16 from Haarlem to the south and the provincial road 1 through the dunes north of the Hague) (see Table 11.2.4). Finally, the flow pattern obtained by the four steps looks somewhat better balanced than that obtained by the three steps. In conclusion, on the basis of the things discussed above for the computations with SALMOF, it was decided to use *four equal steps*.

11.2.3   Choice of the Magnitude of Δx in SALMOF

In Section 5.7.3, the sensitivity of the results for the magnitude of Δx in the marginal objective function has also been investigated. It has been shown that in any case for a steeply increasing convex objective function a large value for Δx is better than a small value. This is owed to the fact that the larger the assigned flow the more serious the effect of an error is. The best results were obtained when the value of Δx approximately equalled the largest assigned value of the flow.

Working out the results of Section 5.7.3 for the objective function, the network and the trip-matrix discussed here produces the following considerations:

(a)   It is very hard to say how large the large flows assigned for the different roads will be. Generally it can be stated that a wide highway will have a flow of five to six thousand p.c.u. in each direction in the average evening peak hour. So the application of four equal steps of value from 1,250 to 1,500 for Δx would seem sensible with regard to the magnitude of the flow assigned.

(b)   The larger Δx, the earlier is the response to the dimension constraints. This is also clear from figure 11.2.2 which gives the value of the marginal objective function as a function of x for different values for Δx.

The situation must be avoided where many roads are overloaded to such an extent that the dimension constraints are approached or 'exceeded'. Not only will this cause too little investment at the SALMOF phase, but also the 'error' will remain uncorrected in the DESCASS/OPTADJ phase. By applying a higher value for Δx, this overloading is avoided. The price we have to pay is that a part of the potential capacity (often existing roads) is no longer used in the SALMOF phase, even before the constraints have been reached (see also Section 11.2.1). However the latter case is not so serious, because the effect is partly corrected anyway in the DESCASS/OPTADJ phase.

We will illustrate this effect with an example featuring a hypothetical objective function and so on. In Figure 11.2.3 the computation process is given for a network consisting of two roads. The left road 1 is non-existent and may be expanded till it has six lanes in each direction: the right road 2 is an existing non-expandable two-lane road ($c_2^{min} = c_2^{max} = 1,600$). In the

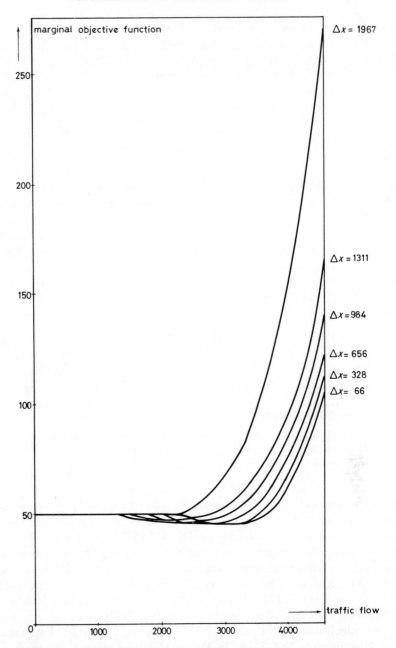

Figure 11.2.2   Marginal objective function in each direction as a function of the traffic flow for a road with maximally 2 × 3 lanes for different values of Δx (N.B. See also the decrease in the marginal objective function for the change of the slope of the investment curve at 2 × 2 lanes)

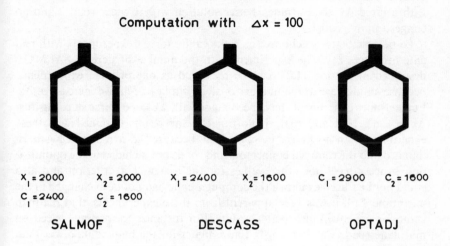

Computation with $\Delta x = 100$

$X_1 = 2000$ $X_2 = 2000$ $X_1 = 2400$ $X_2 = 1600$ $C_1 = 2900$ $C_2 = 1600$
$C_1 = 2400$ $C_2 = 1600$

SALMOF   DESCASS   OPTADJ

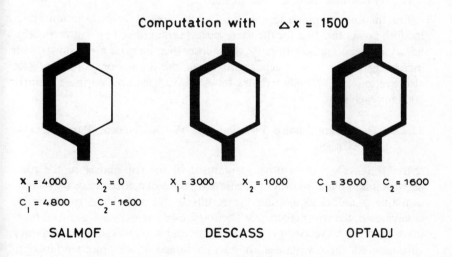

Computation with $\Delta x = 1500$

$X_1 = 4000$ $X_2 = 0$ $X_1 = 3000$ $X_2 = 1000$ $C_1 = 3600$ $C_2 = 1600$
$C_1 = 4800$ $C_2 = 1600$

SALMOF   DESCASS   OPTADJ

Figure 11.2.3   Effect of the magnitude of Δx in SALMOF

first case with a small value for Δx in the final solution there is congestion on road 2 and road 1 has been constructed too narrow. The second process with a large Δx gives a much better solution with a wider road 2 and no congestion on road 1.

To be able better to choose the right values for Δx experiments with real data are necessary. The experiment with the number of steps in SALMOF described in Section 11.2.2 can be reckoned as one of these experiments, because an increase in the number of steps implies a relative increase of Δx. Furthermore experiments have been made with $\Delta x = 66$ (three steps) against $\Delta x = 1967, 1311$ and 1311. Unfortunately the results obtained with these experiments can no longer be compared, because the trip-matrix has been changed and the centroid connectors and roads not included in the optimization process have got other values for the assignment criterion and also could not be traced exactly if the computer code had not been changed in the meantime. Still it was very apparent from these experiments, that the high value for Δx gave much better results. For instance, the average speed on multilane roads was four to six times higher for the high value of Δx than for the low value.

On the basis of the considerations and the results of experiments given above, the following values of Δx have been chosen for the computation described in this chapter:

>   1967 (first step)
>   1311 (second, third and fourth step)

For the final computations of the Dutch Integral Transportation Study the fifth power function for the users' costs (as described in Chapter 9 of this book) has been applied. Because these costs then increase more slowly with increasing traffic flow, when approaching the dimension constraint (see also Section 10.2.2), higher values for Δx have been applied there, namely 2000 for each step.

11.2.4   The Route Choice Function and the Assignment Parameters in DESCASS

In DESCASS a descriptive assignment of the trip-matrix to the road network takes place. In order to achieve this, it is of course necessary to have available a model simulating the results of the individual route choice behaviour of the trip-makers (see Section 2.1.4). Factors such as travel time, travel comfort, travel costs, risk of accidents and so on will act as explanatory variables. All these variables are also contained in the objective function. But in the descriptive assignment these variables are evaluated individually and not socially. Moreover other factors play a part, e.g. the distribution over different routes caused by the size of the transport zones and so on.

As has been said before, the route choice is simulated rather well in a process in which the trip-matrix is assigned to the network by a number of steps (capacity restraint, see Section 2.1.4.2). In this process, the best route is used for every relation at every moment, namely the route with the minimum value for the users' costs as evaluated by the trip-maker.

Taking into account the computation time, it is extremely important that this assignment process is executed in as few steps as possible. This implies that the flows on the roads will increase in large jumps. But, in that case, a fifth or seventh power function is not very well-suited. The rather sudden 'bend' in the users' costs function makes it possible that the flow/capacity ratio can be rather high without much increase in the costs. Also, if the dimension constraint of a road is approached, there is a great chance that this road is assigned more traffic in the next step, so that the road is over-loaded. On the other hand, the users' costs can become so high on an over-loaded road that the traffic is assigned to extreme detours; a result which does not correspond with reality. So the requirement of the small number of steps on the one hand and, on the other, the large number of factors to be taken into account with the simulation of the real route choice behaviour make it impossible to apply directly the users' costs functions discussed in Chapter 9 as the individual route choice criterion.

The object is to achieve a fairly realistic distribution of the traffic flows over the network in as few steps as possible. Extreme overloadings and extreme detours must then be avoided. To this end, in the subroutine DESCASS an assignment is applied according to the routes with minimum users' costs for which these costs increase fairly smoothly with increasing flow/capacity ratio, that is according to a function with the same mathematical shape as the one of Chapter 9 but with a second power for the exponent CCRE:

$$t = \text{TZERO} + \text{CCR}\left(\frac{x}{c}\right)^2$$

For the assignment techniques applied in traffic engineering, the following method is often used to avoid extreme detours. At the 'normal' flow/capacity ratio, travel time is considered as the only decisive cost factor but, at high values for the flow/capacity ratio, other factors which are linearly dependent on the travel distance are considered to be decisive for the route choice. In practice this is realized by using travel time as the criterion up to a certain value of the flow/capacity ratio and by using a constant for the users' costs above that value of the flow/capacity ratio:

$$t = \text{TZERO} + \text{CCR}\left(\frac{x}{c}\right)^{\text{CCRE}} \quad \text{if } \frac{x}{c} \leqslant \sqrt[\text{CCRE}]{\frac{\text{TMAX} - \text{TZERO}}{\text{CCR}}}$$

$$t = \text{TMAX} \quad \text{if } \frac{x}{c} \geqslant \sqrt[\text{CCRE}]{\frac{\text{TMAX} - \text{TZERO}}{\text{CCR}}}$$

This method has not been used in this study.

To get an idea of the sensitivity of the results for the route choice function and to make it possible to choose that function which can best be used, two different route choice functions have been tested:

strong route choice function:

$$t = 0.011 + 0.0512\left(\frac{x}{c}\right)^2 \text{ hour/kilometre}$$

weak route choice function:

$$t = 0.011 + 0.0128\left(\frac{x}{c}\right)^2 \text{ hour/kilometre}$$

The most important macroresults of the total computation (SALMOF with the parameters as discussed before, followed by DESCASS using four steps and OPTADJ) are given in Table 11.2.5.

Table 11.2.5   Effect of the route choice function in DESCASS on the most important macroresults for a test-matrix (in indices)

Function	SALMOF DESCASS strong OPTADJ	SALMOF DESCASS weak OPTADJ
Objective function	100·0	177·2
Investments in two-lane roads	19·2	20·5
Investments in multilane roads	35·2	32·8
Investments in structures	45·6	41·8
Total investments	100·0	95·1
Kilometrage on two-lane roads	31·5	33·5
Kilometrage on multilane roads	68·5	65·0
Total kilometrage	100·0	98·5
Average speed on two-lane roads	100·0	59·1
Average speed on multilane roads	100·0	30·4

Application of the strong route choice function yields a value of the objective function which is only 56 per cent of the value obtained by applying the weak route choice function. This large decrease is totally caused by a better response to possible overloadings. This is also clearly shown by the values for the speed. (Of course these values have been computed applying the *same* speed/flow relationship, namely the seventh power function from Figure 9.2.5.)

The total amount of investments is made a little higher by the application of the strong route choice function (about 5 per cent), as is the total number of kilometres driven (about 1·5 per cent). Moreover a slight shift to the use of wider roads takes place.

The flow patterns resemble each other very much at first sight. Deeper analysis shows, however, that there are still important differences between the results of the strong route choice function compared with those of the weak route choice function. To mention some of them:

1. the new rijksweg 16 east of Rotterdam gets more traffic flow with the strong route choice function;
2. the new rijksweg 3 from Amsterdam to the south gets more traffic flow;
3. the through-going rijksweg 15 from Rotterdam to Twente gets more traffic flow;
4. in Twente the old road passing Delden gets less flow and the highway E8 more;
5. the old road from Den Bosch to Eindhoven gets less flow in favour of the new rijksweg 2; this tendency is continued further south;
6. the old road along the east bank of the Maas in the north of Limburg gets less flow, and the new one along the west bank gets more.

On the basis of the results of the experiment mentioned above, the strong route choice function has been applied in the computations:

$$t = 0.011 + 0.0512\left(\frac{x}{c}\right)^2 \text{ hour/kilometre}$$

This function is illustrated in Figure 11.2.4.

In defining the number of steps for DESCASS, the same considerations play a role as for the route choice function, namely as few steps as possible and a flow pattern as realistic as possible: i.e. no large overloadings where 'parallel' routes are present and no extreme detours. To meet these two— conflicting—requirements as well as possible, four steps have been used. In the first step a 0·1 part of the trip-matrix is assigned to the network. It is necessary to make the first step a small one, because in SALMOF some roads come up with very small dimensions. Because the users' costs on such roads are still low in the first step of DESCASS, they are apt to be soon overloaded. In each of the subsequent three steps a 0·3 part of the trip-matrix is assigned. If we could have used more steps we would have preferred to make the last

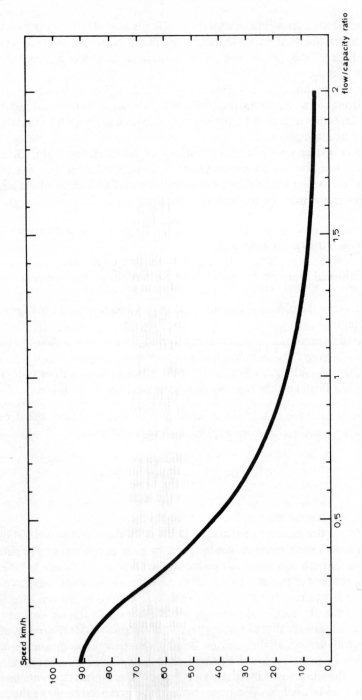

Figure 11.2.4   Route choice function applied in DESCASS

few smaller. However, with a total of four steps, this would have meant a large second step with too great a chance of overloads.

11.3   ILLUSTRATION OF THE OPERATION OF THE METHOD ON SOME PLACES OF THE NETWORK

To get a better idea of the operation of the computation process, we will analyse here the intermediate results after the different steps of SALMOF and after DESCASS/OPTADJ. We will always consider one road or a group of roads, taken from the whole network.

The way in which the traffic is assigned to the roads depends not only on the transport pressure at both ends of the related roads and the characteristics of those roads, but also on the wider situation in the network which sometimes means a very extensive surrounding area. This is also shown by the fact that the total of the flows assigned is different for every step.

11.3.1   The Roads between Utrecht and The Hague and Rotterdam

The 'green heart of Holland' in the centre of the Randstad is cut through in the east–west direction by three major highways: (a) the northern rijksweg 11 going from Maarssen north of Utrecht (connecting with rijksweg 2 from Utrecht to Amsterdam and then rijksweg 27 from Utrecht to Hilversum), proceeding south of Alphen and then connecting with rijksweg 4B, rijksweg 4 and rijksweg 44 all from Amsterdam to The Hague; (b) the existing rijksweg 12 proceeding from Arnhem directly south of Utrecht to Gouda, where this road is split into a highway to The Hague and a highway to Rotterdam; and finally (c) the southern rijksweg 24 connecting with rijksweg 2 from Utrecht to Den Bosch and going *via* IJsselstein to the northern part of the circular road around Rotterdam. It is assumed that rijksweg 11 and rijksweg 24 will not yet exist as highways in 1975; there are only secondary two-lane roads in the vicinity of their future paths. The existing rijksweg 12 is not allowed to be expanded beyond 2×4 lanes on the section Utrecht–Gouda, as the general constraint of 2×6 lanes holds for the other two roads only. The three roads each belong to the category of a road in a rural area in the Randstad. Though rijksweg 24 and rijksweg 11 have the same cost characteristics, there is still a large cost difference. Rijksweg 11 must be constructed fully as a highway, while rijksweg 24 is connected to the existing rijksweg 12 west of Gouda (in 1975 2×3 lanes, and expandable up to 2×6 lanes). So rijksweg 24 is 'cheaper' than rijksweg 11. The dimension constraints of the three roads on the section east of Gouda are given in Table 11.3.1.

We will now consider the traffic flows and the number of lanes in the three roads on the sections east of Gouda, after each step of SALMOF and after the last step of DESCASS with the number of lanes as defined in OPTADJ.

Table 11.3.1   Dimension constraints of the roads between
Utrecht and The Hague and Rotterdam on the sections east of
Gouda

Road	Number of lanes in 1975	Maximum number of lanes
Rijksweg 11	1×2	2×6
Rijksweg 12	2×3	2×4
Rijksweg 24	1×2	2×6

These flows are given in Table 11.3.2 and in Figure 11.3.1. Figure 11.3.1(a)
also gives the flow-pattern obtained by a descriptive assignment to a network,
in which all roads are present with maximal dimension.

In the first step of SALMOF the existing rijksweg 12 is almost the only
highway used. On this road no investments are now necessary. Moreover
it is indicated by an assignment according to the shortest distance (the
results of this all-or-nothing assignment are not discussed further here)
that the main transport potentials are also situated here. Rijksweg 24 gets
more traffic than rijksweg 11 as can be explained by the fact that rijksweg
24 is connected with the existing rijksweg 12 east of Gouda and thus is
cheaper.

In the second step rijksweg 12 must be expanded from 2×3 to 2×4
lanes. However, this is cheaper than the expansion of two-lane roads as
would be necessary for rijksweg 11 and rijksweg 24. So we see that the largest
increase of flow occurs on rijksweg 12. Still rijksweg 24 gets a larger increase
of flow than in the first step, whereas the increase in flow on rijksweg 11
equals that of the first step.

In the third step, for rijksweg 12 the dimension constraint can be felt
quite well, especially in the west–east direction, where the traffic flow is
largest. In this direction there is therefore no increase in flow, while the
flow increases very little in the opposite direction. The increase in flow
appears almost totally on the two other highways and mainly on rijksweg 24.

In the fourth and last step, the computation proceeds almost in the same
way as in the third step.

In the next phase of the computation process, the flow pattern is defined
in a descriptive assignment. We hardly see a shift in the distribution of the
flows over the three highways. The effect of again using non-expandable
two-lane roads described in Section 11.2.1 is shown very clearly (see the
roads south of rijksweg 24 and between rijksweg 11 and rijksweg 12 on
Figure 11.3.1). The total flow decreases considerably in favour of the two-
lane roads to the north and to the south. As a result, the overloading of
rijksweg 12, which emerged in SALMOF, also vanishes.

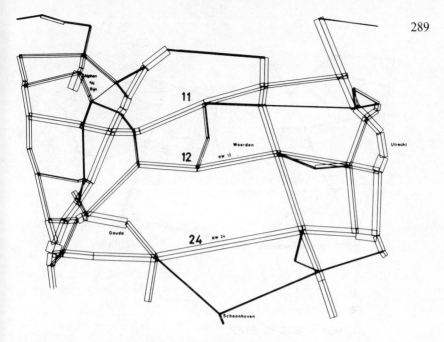

DESCRIPTIVE ASSIGNMENT TO UNLIMITED NETWORK

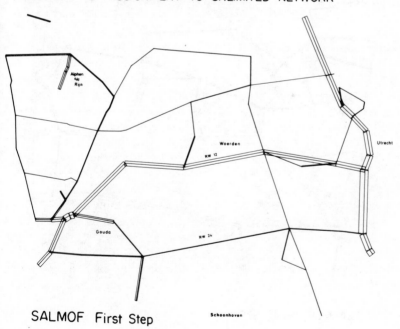

SALMOF First Step

Figure 11.3.1(a)   Illustration of the method for the roads between Utrecht and
The Hague and Rotterdam

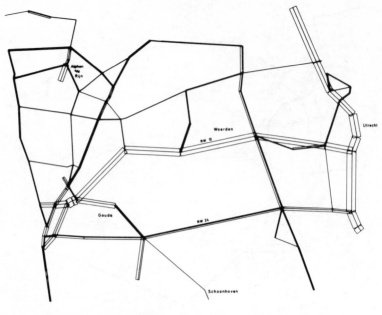

SALMOF SECOND STEP

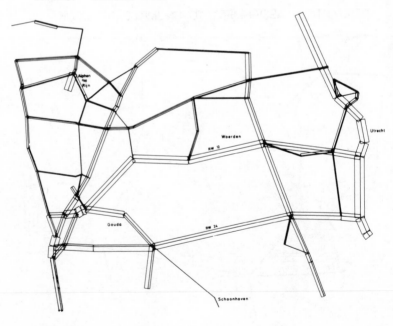

SALMOF THIRD STEP

Figure 11.3.1(b)   Illustration of the method for the roads between Utrecht and
The Hague and Rotterdam

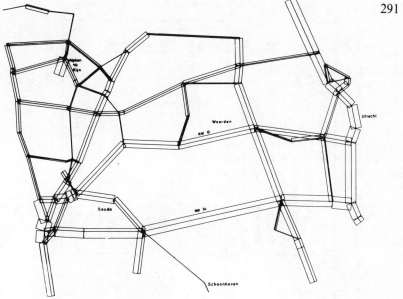

SALMOF LAST STEP

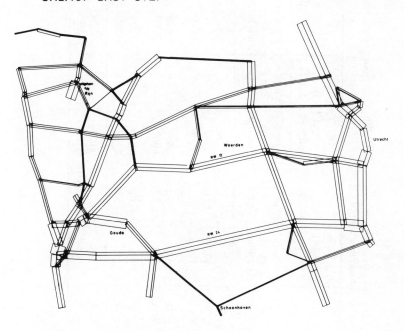

FINAL SOLUTION

Figure 11.3.1(c)   Illustration of the method for the roads between Utrecht and The Hague and Rotterdam

Table 11.3.2   Traffic flows in p.c.u. in the average evening peak hour and the numbers of lanes for the roads between Utrecht and The Hague and Rotterdam on the sections east of Gouda

Road	SALMOF after first step		SALMOF after second step		SALMOF after third step		SALMOF after last step		SALMOF/DESCASS/OPTADJ final solution	
	Flow in evening peak hour	Number of lanes	Flow in evening peak hour	Number of lanes	Flow in evening peak hour	Number of lanes	Flow in evening peak hour	Number of lanes	Flow in evening peak hour	Number of lanes
Rijksweg 11										
E–W[a]	172	1·0	312	1·0	1,317	1·3	2,215	1·9	2,338	2·0
W–E	155	1·0	279	1·0	1,416	1·4	2,695	2·6	2,974	2·8
Rijksweg 12										
E–W	2,636	3·0	4,449	4·0	4,628	4·0	4,820	4·0	4,428	4·0
W–E	3,224	3·1	5,067	4·0	5,067	4·0	5,067	4·0	4,218	4·0
Rijksweg 24										
E–W	576	1·0	1,341	1·3	3,424	3·2	6,132	5·2	5,072	4·4
W–E	288	1·0	1,580	1·5	4,597	4·0	6,980	5·9	4,976	4·3
Total E–W	3,384		6,102		9,369		13,167		11,838	
W–E	3,667		6,926		11,080		14,742		12,168	

[a] E–W : from east to west

11.3.2 The Road Network North of The Hague

From The Hague to the north, six roads are of importance. These are, from west to east : (a) the provincial road 1 through the dunes, continuing to Haarlem; (b) rijksweg 44 through Wassenaar, proceeding west of Leiden and then connecting with rijksweg 4 to Amsterdam; (c) the provincial road '16 *bis*' from The Hague to Oegstgeest and Leiden (the so-called 'Leidse Baan'); (d) the secondary road passing by Leidschendam and Voorschoten (the Veursestraat); (e) rijksweg 4, proceeding from The Hague to Amsterdam, in the south connecting with rijksweg 13 from Rotterdam; and finally (f) rijksweg 4B from rijksweg 16 going parallel to rijksweg 4.

The numbers of lanes which it is assumed will be present in 1975 and the maximum number of lanes are given in Table 11.3.3.

Table 11.3.3 Dimension constraints of the roads north of The Hague

Road	Number of lanes in 1965	Maximum number of lanes
Provincial road 1	—	2×6
Rijksweg 44	2×2	2×3
Leidse Baan	2×2	2×3
Veursestraat	1×2	1×2
Rijksweg 4	2×2	2×6
Rijksweg 4B	—	2×6

The investment characteristics are identical for the four western roads, each belonging to the category of a road in an urban area in the Randstad, just as rijksweg 4 and rijksweg 4B each belong to the category of a road in a rural area in the Randstad.

The traffic flows in the average evening peak hour and the numbers of lanes after each step of SALMOF and after the last step of DESCASS (followed by OPTADJ) are given in Table 11.3.4 and Figure 11.3.2.

In the first step of SALMOF, the new roads for which investments are necessary, and also the non-expandable two-lane road, are hardly loaded if at all. Of the two roads passing by Wassenaar (rijksweg 44 and the Leidse Baan) rijksweg 44 gets almost all the traffic. This must be explained by the fact that the transport potential is mainly situated on the western side. In an assignment according to the shortest distance, most traffic even appears to be assigned to the provincial road 1 through the dunes furthest west.

In the second step, rijksweg 44 and rijksweg 4 must be expanded. The costs related to this expansion do not for every relation outweigh the detour costs related to the use of the existing Leidse Baan. A lot of traffic is therefore assigned to the latter road. The new roads and the non-expandable one are still unattractive.

Table 11.3.4   Traffic flows in p.c.u. in the average evening peak hour and the numbers of lanes for the roads north of The Hague

Road	SALMOF after first step		SALMOF after second step		SALMOF after third step		SALMOF after last step		SALMOF/DESCASS/OPTADJ final solution	
	Flow in evening peak hour	Number of lanes	Flow in evening peak hour	Number of lanes	Flow in evening peak hour	Number of lanes	Flow in evening peak hour	Number of lanes	Flow in evening peak hour	Number of lanes
Provincial road 1										
N–S[a]	25	0·0	117	0·1	386	0·4	917	1·0[b]	2,651	2·4
S–N	82	0·1	185	0·2	1,438	1·3	3,297	3·1	3,874	3·6
Rijksweg 44										
N–S	1,191	2·0	1,686	2·0	2,082	2·0	2,352	2·1	2,219	2·1
S–N	2,184	2·0	2,636	2·4	3,218	2·9	3,218	2·9	2,955	2·7
Leidse Baan										
N–S	167	2·0	2,102	2·0	2,827	2·6	3,054	2·8	1,161	2·0
S–N	333	2·0	3,654	3·0	3,654	3·0	3,654	3·0	2,713	2·5
Veursestraat										
N–S	0	1·0	0	1·0	0	1·0	0	1·0	820	1·0
S–N	0	1·0	0	1·0	67	1·0	196	1·0	1,054	1·0
Rijksweg 4										
N–S	2,171	2·0	2,476	2·3	3,712	3·4	4,460	3·9	2,767	2·5
S–N	2,953	2·8	3,772	3·5	5,909	4·9	6,716	5·5	5,507	4·6
Rijksweg 4B										
N–S	133	0·1	229	0·2	324	0·3	1,512	1·4	1,647	1·5
S–N	117	0·1	249	0·3	555	0·6	2,367	2·2	2,409	2·2
Total N–S	3,687		6,610		9,331		12,295		11,265	
S–N	5,669		10,496		14,841		19,448		18,512	

[a] N–S: from north to south

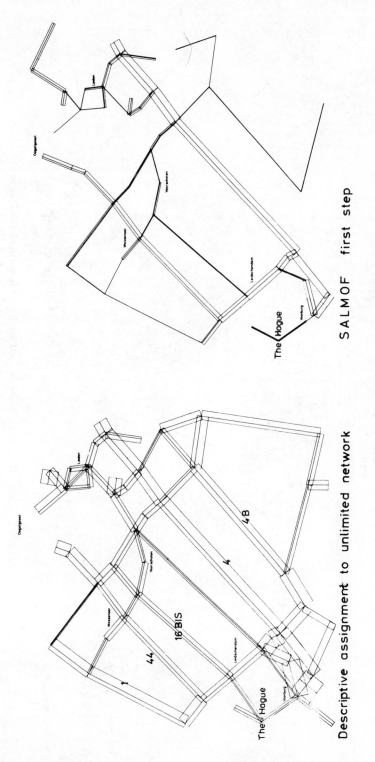

Descriptive assignment to unlimited network

SALMOF first step

Figure 11.3.2(a)   Illustration of the method for the roads north of the Hague

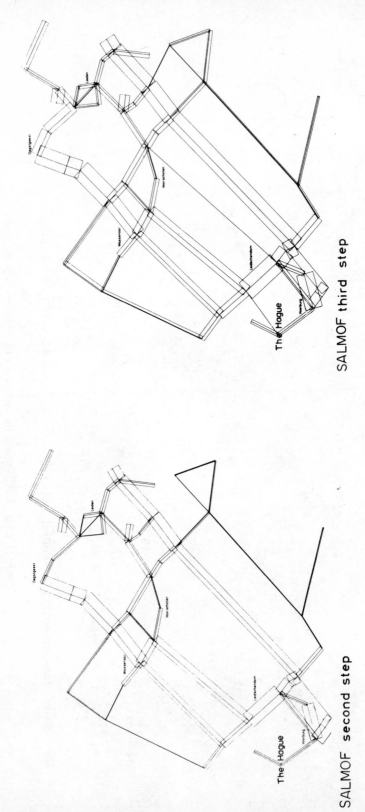

SALMOF third step

SALMOF second step

Figure 11.3.2(b) Illustration of the method for the roads north of The Hague

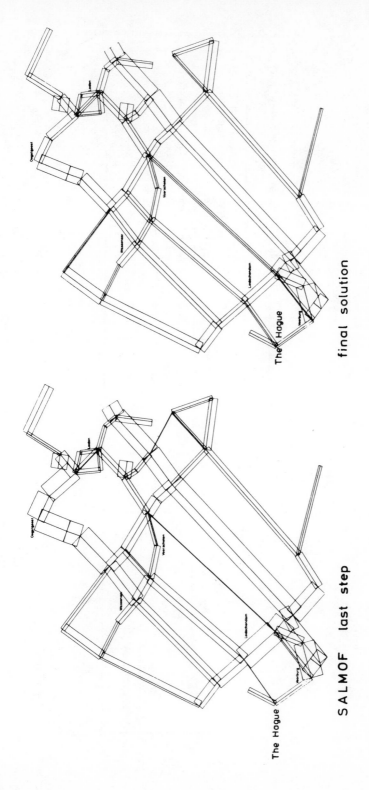

SALMOF   last step

final solution

Figure 11.3.2(c)   Illustration of the method for the roads north of The Hague

Table 11.3.5   Traffic flows in p.c.u. in the average evening peak hour and the numbers of lanes for rijksweg 58 between Bergen op Zoom and Goes

Road	SALMOF after first step		SALMOF after second step		SALMOF after third step		SALMOF after last step		SALMOF/DESCASS/OPTADJ final solution	
	Flow in evening peak hour	Number of lanes	Flow in evening peak hour	Number of lanes	Flow in evening peak hour	Number of lanes	Flow in evening peak hour	Number of lanes	Flow in evening peak hour	Number of lanes
Rijksweg 58 E–W	712	2·0	1,302	2·0	1,893	2·0	2,786	2·3	2,535	2·1
W–E	534	2·0	1,037	2·0	1,482	2·0	1,973	2·0	1,872	2·0

At the beginning of the third step, the Leidse Baan and rijksweg 44 are heavily loaded especially in the direction from The Hague. The dimension constraints of 2×3 lanes cause traffic to be assigned to the provincial road 1 through the dunes, again especially the traffic from The Hague. Even the overloaded Leidse Baan gets no traffic from the south. In the opposite direction, towards The Hague, the existing roads are expanded further to 3 lanes. Also, rijksweg 4 is expanded further. At the end of the second step, the flow on this rijksweg 4 is so high that an expansion beyond 2×4 lanes must already be taken into consideration. This fact results in a small increase in the increase in flow on rijksweg 4B from the south. Finally, the Veursestraat gets some traffic flow as well, though very little.

At the beginning of the fourth step, the two existing roads through Wassenaar have been so heavily overloaded with traffic coming from The Hague that they get no traffic in this direction. This traffic is now assigned to the provincial road 1 through the dunes. Rijksweg 4 must clearly now be expanded beyond 2×4 lanes. The costs involved in this expansion almost equal the costs for constructing a new highway. So we see a distribution of the traffic assigned over rijksweg 4 and rijksweg 4B.

For the descriptive assignment, only the users' costs are relevant in defining the routes. So we see a shift in the westerly direction, The provincial road 1, especially, gets more traffic* and the Leidse Baan less. Next we see very clearly the process discussed in Section 11.2.1, that existing non-expandable roads are not assigned in SALMOF but are assigned in DES-CASS. The Veursestraat gets a reasonable traffic flow now and, because of this, the flow on rijksweg 4 decreases.

11.3.3 Rijksweg 58 between Bergen op Zoom and Goes

If there are not many alternatives for a certain road, the flows assigned in the different steps of SALMOF will be equal to each other and there will be no large differences between the final flow of SALMOF and that of DES-CASS. An example of such a road is provided by rijksweg 58 between Bergen op Zoom and Goes in Zeeland. The traffic flows and dimensions are given in Table 11.3.5 and the equal flows assigned at the different steps are shown.

11.3.4 The Afsluitdijk

The Afsluitdijk, the 30 kilometre-long connection from North Holland to the northern provinces of the Netherlands across the sea, seems rather isolated at first sight. Moreover the Afsluitdijk is a 2×2 lane road, which

---

* In the computation process discussed here, the dimensions of the road in both directions are set equal to the largest dimension after SALMOF but before DESCASS. This explains the large flow on the provincial road 1 in the north–south direction. In the final computations of the Dutch Integral Transportation Study, this process of setting equal has not taken place.

Table 11.3.6   Traffic flows in p.c.u. in the average evening peak hour and the numbers of lanes for the Afsluitdijk

Road	SALMOF after first step		SALMOF after second step		SALMOF after third step		SALMOF after last step		SALMOF/DESCASS/ OPTADJ final solution	
	Flow in evening peak hour	Number of lanes	Flow in evening peak hour	Number of lanes	Flow in evening peak hour	Number of lanes	Flow in evening peak hour	Number of lanes	Flow in evening peak hour	Number of lanes
Afsluitdijk E–W	891	2·0	1,352	2·0	1,779	2·0	2,468	2·4	2,071	2·0
W–E	799	2·0	1,193	2·0	1,552	2·0	2,297	2·2	1,872	2·0

limensions seem quite adequate. So one would expect the same regular
ncrease of flow. However, there exist alternatives for the Afsluitdijk. This
like is, of course, (largely) used for long-distance trips between the Randstad
and the northern provinces of Groningen and Friesland. But for these very
trips the routes through the new Flevopolders are attractive alternatives
and the routes are defined by the whole situation of the network in North
Iolland on the one hand and all eastern and northern provinces on the other.

In the first and the last step of SALMOF, the flows assigned are almost
wice as large as in the two middle steps (see Table 11.3.6). In the first step of
ALMOF, the use of non-expandable two-lane roads is almost excluded
nd, in the last step, many roads are unattractive due to impending over-
oading or high expansion investments. This will have to explain the irregular
ssignment to the Afsluitdijk.

1.4 COMPARISON OF THE RESULTS OBTAINED BY THE STEP-
ISE ASSIGNMENT ACCORDING TO THE LEAST MARGINAL
BJECTIVE FUNCTION WITH THOSE OBTAINED BY AN
PTIMAL ADJUSTMENT OF THE DIMENSIONS TO THE
TRAFFIC FLOWS

To get an idea of the (useful) effect of the application of the stepwise
ssignment according to the least marginal objective function, we compare
he results obtained by this method with the road network and flow pattern
btained in another way. As a basis for the comparison we choose the first
tep of the continuous optimal adjustment of the dimensions to the traffic
ows, a heuristic procedure described in Section 4.4.2.1 which resembles
real decision-making process. In order to achieve this, the trip-matrix
assigned first to a network in which all roads have their maximal dimension
he subroutine DESCASS). Then the optimal number of lanes for every
oad is defined for the traffic flow on that road (the subroutine OPTADJ).

As was said in Section 11.2.1, the stepwise assignment according to the
ast marginal objective function is followed in our computation process
y a descriptive assignment of the trip-matrix to the network defined earlier
d by an optimal adjustment of the dimensions of the roads to the traffic
ows obtained through the descriptive assignment (SALMOF/DESCASS/
PTADJ). For the sake of simplicity we will call this total computation
ocess the optimization process in the rest of this section.

The results of these two computations have already been given briefly
Section 11.2.1 (Table 11.2.1 first and third columns) but here we will go
to further details. The most important macroresults are given in Table
.4.1.

Comparing the macroresults with each other we see the following:
the optimization process gives a lower value for the objective function;

Table 11.4.1 Macroresults obtained by an optimal adjustment of the dimensions to the traffic flows and by the stepwise assignment according to the least marginal objective function followed by a descriptive assignment and an optimal adjustment

Function	Optimal adjustment (DESCASS/OPTADJ)	Optimization process (SALMOF/DESCASS/OPTADJ)
Objective function (thousand millions of Dutch guilders)	132·5	115·4
Investments in two-lane roads	3·4	1·8
Investments in multilane roads	13·2	12·1
Investments in structures	15·4	14·6
Total investments (thousand millions of Dutch guilders)	32·0	28·5
Kilometrage on two-lane roads	4·15	4·63
Kilometrage on multi-lane roads	23·29	22·92
Total kilometrage (million kilometres in the average evening peak hour)	27·44	27·56
Average speed on two-lane roads	26·8	44·7
Average speed on multi-lane roads (kilometres/hour)	83·0	83·2

(b) in the optimization process hardly any more kilometres must be driven (an increase of about 0·4 per cent);
(c) in the optimization process there are less congestions;
(d) the amount of investment is considerably lower for the optimization process (12 per cent; 3·5 thousand million Dutch guilders in absolute monetary terms).

So we obtained better results using the optimization process than using the optimal adjustment.

It must be noted, however, that the trip-matrix used here consisted of so many trips that in some parts of the country for both methods all possible road infrastructure had to be used. For trip-matrices with less trips, the difference in results between the two methods would have been larger.

The better results of the optimization process compared with those of the optimal adjustment can be explained by the following factors:

(a) In the optimization process the existing roads are better used.

(b) In the optimization process the construction and use of 'expensive' roads is avoided.

(c) In the optimization process a certain choice is made between potential roads—if possible. This has two advantages:

    (1) the 'cheapest' road will always be chosen;

    (2) where the objective function is concave (starting area) the construction and use of one road gives a lower value for the objective function than that of two roads.

(d) The higher value for the average speed in the optimization process looks difficult to explain at first. In defining the flow pattern according to the descriptive assignment, for both methods the same assignment parameters have been used in the subroutine DESCASS. At the assignment to the maximal road network, because all roads expandable to 2×6 lanes are present with their fullest possible dimension, one would expect heavy flows on those roads and, related to that, a high average speed after the optimal adjustment of the dimensions to the flows. For the dimension-constrained roads are very unattractive in the descriptive assignment to the maximal road network and on those very roads congestions later arise. The still higher average speed obtained by the optimization process can be explained as follows:

    (1) the roads with low investment costs are used most in the optimization process; on these roads a high speed is possible;

    (2) during the optimization process a more coherent road network is composed; that means that a two-lane road followed by a 2×6-lane road is hardly attractive, whereas with the pure descriptive assignment such a pattern of use could be a good solution.

It is interesting to see that the investments in two-lane roads are much higher in the optimal adjustment. This indicates a large number of small roads.

The differences indicated above between the optimal adjustment and the optimization process can best be seen in the resulting networks and flow patterns. We will discuss below some parts of the network where the differences between the results of the optimal adjustment and the optimization process are large. The parts mentioned below are only a selection from the full network. Of course the differences also occur elsewhere.

(a) From Haarlem to Amsterdam three main highways go to The Hague and Rotterdam: the existing rijksweg 4, the new rijksweg 16 and the new rijksweg 3. In the optimization process the existing rijksweg is used much more and the new highways less than in the pure descriptive assignment (see Table 11.4.2).

Table 11.4.2   Traffic flows in p.c.u. in the average evening peak hour on the three main highways from Haarlem and Amsterdam to The Hague and Rotterdam (on the line Lisse–Leimuiden–Wilnis)

Road		Optimal adjustment	Optimization process
Rijksweg 16	N–S	4,740	3,180
	S–N	4,990	3,500
Rijksweg 4	N–S	4,780	5,600
	S–N	3,780	5,890
Rijksweg 3	N–S	5,180	4,750
	S–N	5,500	3,850

(b)   The three highways from Utrecht to The Hague and Rotterdam have also been discussed in Section 11.3.1 in order to illustrate the operation of the method. Here, in the comparison of the optimal adjustment and the optimization process, we see that the pure descriptive assignment causes a more-or-less equal distribution over the three roads, whereas in the optimization process the most expensive highway (rijksweg 11) is avoided in favour of the cheapest highway (see Table 11.4.3).

Table 11.4.3   Traffic flows in p.c.u. in the average evening peak hour on the highways from Utrecht to The Hague and Rotterdam

Road		Optimal adjustment	Optimization process
Rijksweg 11	E–W	4,010	2,340
	W–E	3,800	2,970
Rijksweg 12	E–W	3,150	4,430
	W–E	4,120	4,220
Rijksweg 24	E–W	5,010	5,070
	W–E	4,620	4,980

(c)   In the Flevopolders, the operation of the optimization process is shown clearly (see tables 11.4.4 and 11.4.5). In the pure descriptive assignment all roads are almost equally loaded, whereas in the optimization process a definite choice is clearly made. Moreover, in the optimization process the existing non-expandable two-lane road along the Oostvaardersdiep is used to better advantage.

(d)   In the pure descriptive assignment, the existing road from Venlo to Well is not used at all, whereas the new parallel rijksweg 73 is heavily loaded. In the optimization process the existing road is well used.

Table 11.4.4 Traffic flows in p.c.u. in the average evening peak hour on the roads through the Flevopolders from south-east to north-west

Road		Optimal adjustment	Optimization process
Huizen–Pampus	E–W	530	0
	W–E	740	0
Nijkerk–Marken	E–W	3,800	5,660
	W–E	3,800	6,200
Voorthuizen–Warder	E–W	2,090	1,250
	W–E	2,350	1,360
From Harderwijk	E–W	900	1,090
	W–E	1,660	1,130

Table 11.4.5 Traffic flows in p.c.u. in the average evening peak hour on the roads through the Flevopolders from north-east to south-west

Road		Optimal adjustment	Optimization process
Through the Markerwaard	N–S	2,140	1,510
	S–N	2,840	1,730
Along the Oostvaardersdiep	N–S	1,180	1,320
	S–N	480	1,440
Rijksweg 6	N–S	2,570	4,910
	S–N	2,400	5,030
To Huizen	N–S	3,260	1,850
	S–N	2,860	1,710

(e) In the optimal adjustment an expensive road must be constructed from Milsbeek near Gennep to Arnhem. In the optimization process this road need only be constructed for a short distance and with fewer lanes.

(f) In the Delta-area (the islands and peninsulas in Zeeland and South Holland), the effect of the optimization process is shown very clearly indeed. Many expensive roads loaded at the pure descriptive assignment and so 'constructed' at the optimal adjustment are not used in the optimization process. This is the case for the following roads among others:

(1) Goes–Terneuzen; about 20 km with a tunnel under the Westerschelde of about 7 km;

(2) Bergen op Zoom–Ossendrecht–Hulst; about 35 km with a tunnel under the Westerschelde of about 15 km;

(3) Kruiningen–Tholen; about 15 km with a tunnel or a bridge crossing the Oosterschelde of about 8 km;

(4)   Zierikzee–Hellevoetsluis—Zwartewaal; about 35 km with two long tunnels or bridges crossing the Grevelingen and the Haringvliet;

(5)   Abbenbroek–Middelsluis; about 20 km with a bridge crossing the Spui.

The non-construction of these five roads implies a large difference in investment costs.

11.5 THE RESULTS OF THE COMPUTATION AND THEIR INTERPRETATION

The computation results in values for the objective function and for the different parts of it and also in values for the decision variables. The main results are:

(a)   the objective function;
(b)   the amount of investments;
(c)   the travel times;
(d)   the road network and the flow pattern on it.

Though the value of the objective function may be the most relevant variable in an optimization process, this value is very difficult to interpret and, on its own, is not in our case very interesting. This value becomes very important indeed if it can be compared with the total social costs for other sectors of society, such as health care, education, defence and so on, or with the total costs of transportation taking other assumptions for basic data, such as the spatial structure, the distribution over time of the traffic, the modal split and so on.

For many people, the amount of investment is a more interesting variable, especially when the distribution of the means available over the different sectors is at question. However, one must be very cautious in interpreting the amount of investment determined in the computation process. In the Dutch Integral Transportation Study, the amount of investment equals the construction costs necessary to build the network resulting from the computation, starting from the network which it is assumed will be present in 1975. In reality, the network will be constructed in stages. The construction costs necessary for this will be (considerably) higher. Next the network is defined in such a way that it will be optimal for the traffic of the related year (see Chapter 8). This implies that the investments have to be made in the preceding years. Moreover, in finally defining the investment scheme, factors may play a role which are not included in the objective function, e.g. a desire for equal distribution of construction activities over time and over the country and so on. The amount of investment required then will differ, of course, from the construction costs determined in the optimization process.

The optimization process results in a network of roads with a fractional number of lanes (see Section 10.4) generally different for each direction. Such a network with related construction costs is, of course, rather unrealistic. Another possibility is to set the number of lanes for both directions equal to the number of lanes for the more heavily-loaded direction. In some cases, this will not be necessary because it is, of course, possible to use some lanes in one direction at certain times and in the opposite direction at others. The difference in construction costs for the networks with unequal and equal numbers of lanes in each direction is about 20 per cent, for the situation discussed in this chapter (taking the larger of the two dimensions for the two directions as the dimension for both directions with equal dimensions).

In Section 10.4 the 'rounding off' of a fractional number of lanes into an integral number has already been discussed. There the criterion of the minimization of the objective function for one road was applied. Depending on the different assumptions and the dimension level, this minimization resulted in a desired 'rounding off' for a lane at a value 0·1 to 0·5 part of a lane. In a network, the situation is somewhat more complicated. Three roads with 1·25 lanes may be rounded off to three two-lane roads or to one two-lane and two one-lane roads. In this last case, the rounding off will also imply another choice of routes in reality and therefore other values for the users' costs as well.

In Sections 5.4.2 and 10.4 it has been pointed out that, by applying integral numbers of lanes, the objective function does not everywhere possess the required convex shape. So it is not easily possible to apply integral numbers of lanes immediately. Defining the desired integral number of lanes and the related investment costs after the computation in fractional numbers is, however, a difficult and rather arbitrary affair. On the other hand, the impact on the total amount of investments can be quite heavy: 'rounding off' at a quarter of a lane

$$c := c_n \text{ for } c_n - 0\cdot75c_1 \leqslant c \leqslant c_n + 0\cdot25c_1$$

implies 15 per cent more investment costs than without this 'rounding off'. In decisions concerning road networks, this uncertainty does no harm because in reality the roads will not all be constructed together by a certain date.

However, considering the amount of investment on its own, one must take this margin well into account. Therefore, in the Dutch Integral Transportation Study, the final amounts of investment have been given in many different ways.

Finally, it is obvious that the amounts of investment are no better estimated than the construction cost functions describe the real construction costs.

The travel times form an important indicator for the quality of traffic operation on the road network (see Section 9.2.2). The average speed can be computed in different ways:

$$\bar{v} = \frac{\sum\limits_{ij \in L} x_{ij} l_{ij}}{\sum\limits_{ij \in L} x_{ij} z_{ij}} \tag{11.5.1}$$

$$\bar{v}' = \frac{\sum\limits_{ij \in L} l_{ij}}{\sum\limits_{ij \in L} z_{ij}} \tag{11.5.2}$$

Relationship (11.5.1) gives the average speed weighted with the lengths of and the traffic flows on the different links; i.e., the average speed in its importance to all the trip-makers, which is the most relevant value. Relationship (11.5.2) gives the average speed weighted only with the lengths of the links. This average value may be of relevance if one is interested only in the situation on the roads themselves.

The average speed can be quite heavily influenced by serious congestion on a few roads. To eliminate this effect, it may be important to compute the median speed or to assume a certain minimum value for the speed in the computation of the average speed, so:

if
$$\frac{x_{ij}}{c_{ij}} \geqslant \sqrt[\text{CCRE}]{\frac{\text{TMAX} - \text{TZERO}}{\text{CCR}}}$$

then:

$$z_{ij} = l_{ij}\text{TMAX}$$

The absolute value of the (average) speed is of less interest than the comparison with the 'optimal' speed, that is with:

$$\text{TZERO} + \text{CCR}\left(\frac{A}{D1 . \text{CCRE}}\right)^{\frac{\text{CCRE}}{1 + \text{CCRE}}}$$

(see Section 10.1.3).

Higher speeds can occur in those places where existing road infrastructure is present to a large extent. Lower speeds occur where the dimension constraints are approached or 'exceeded'. Such an approach and/or 'exceeding' can be caused by different facts:

(a) Though there are sufficient 'parallel' routes, a dimension-constrained road is still overloaded owing to a wrong choice of the number of steps, the magnitude of the parts assigned and/or Δx (see Section 11.2.2 and 11.2.3). In this category we number also those cases from which the overloading can be removed by a slight change of the input network with its constraints.

(b) Though 'parallel' routes exist, the extra users' costs and possibly investments caused by the necessary detours do not outweigh the congestion costs on the overloaded road.

(c) There are no other possibilities; the maximally-offered road infrastructure is insufficient.

The low speed computed for congestion of the first category (a) is not very important. The network computed will have to be adjusted a little in some places and the total amount of investment will thereby increase a little. The categories (b) and (c) are more interesting. Assuming that the input maximal network really gives the maximal possibilities (see Section 10.1.4 for the interpretation of the maximum dimension), serious problems arise here. In the method applied, the trip-matrix stays constant (see Chapter 8). In reality this will not be so in these very places. A change will occur in the expected modal choice, the expected transport distribution, the expected transport production up to and including the expected land use pattern. These effects and their social evaluation have not been included in the optimization process and/or the objective function. However, for the interpretation of the results they must be taken into account.

For the optimization discussed in this chapter, some congestions of the category (a) occur, especially in existing non-expandable two-lane roads. On this type of road, some congestions of the category (b) also occur. Direct connections between population nuclei are usually concerned. Congestions of the category (c) are the most interesting of all. These also occur on large (up to 12-lane) highways. These congestions are almost always caused by the traffic to and from large towns and other very important concentrations of employment (for instance the blast-furnace area).

The most illustrative and, in our opinion, also the most important results are provided by the drawings (Calcomp plots) of road networks incorporating the dimensions of or the traffic flows on the roads (the decision variables). In Figure 11.5.1, the reduction of such a plot is given. The normal size is about 70 × 55 cm. There are many possible plots to draw, all of which can be of importance. To mention some of them:

1. flow patterns;
2. plots showing the numbers of lanes;
3. plots showing the investments;
4. plots showing the speeds;
5. plots showing the most important congestions;
6. intermediate results of the computations (for instance the flow patterns after the different steps of SALMOF);
7. plots showing the through-going and different turning flows for an intersection (the so-called 'turning volumes');
8. plots showing which relations use a certain link (the so-called 'selected link');

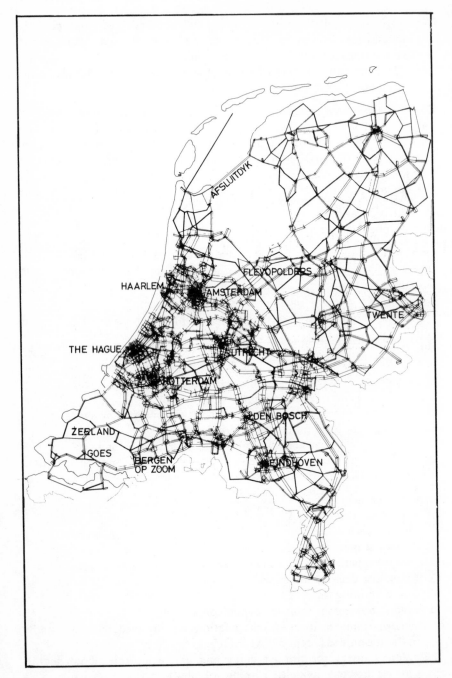

Figure 11.5.1   Plot of the traffic flows on the road network of the Netherlands defined with the stepwise assignment according to the least marginal objective function (test computation)

9. plots showing only the traffic to or from certain towns or regions;
10. plots showing only the traffic flows with a trip-length above or below a certain distance;
11. assignments to certain networks according to the shortest distance or defined with a descriptive model.

This list is far from exhaustive and can be readily extended with a little imagination.

11.6 POLICY IMPLICATIONS

In Chapter 8, the position of the minimization of the social costs for a given matrix of car traffic in the whole of the transportation study and for the definition of spatial structure has been extensively dealt with. In the interpretation and evaluation of the results, it is important to keep this position in mind. Moreover, it is necessary to be continuously aware of which objective function, which decision variables and which constraints have been applied and what assumptions underlie them.

Generally speaking, we may state that the Dutch Integral Transportation Study presents the situation which would arise if on the one hand trip-makers continued to respond to circumstances as they do now (i.e. the transport forecasting model) and if on the other hand the government continued its present policy as far as the whole of physical planning and transportation planning is concerned (the given spatial structure, the objective function and the absence of any radical measure to regulate the transportation system).

Now, the optimization process is meant to yield 'the best solution'. Nevertheless, the question must be asked whether or not this 'best' solution is actually desirable. The best solution in the sense of our optimization process would be the one with the lowest value for the objective function and would be achieved by choosing the proper values for the decision variables within the possibilities deliminated by the constraints. Whether such a solution is actually a desirable one falls outside the problem as stated here. It can and must be considered on a higher level. There the question is (again) asked whether the right objective function, the right constraints and/or the right decision variables have been used. The objective function used for the optimization of the road network may (appear to) be contrary to a social objective on a higher level. It is also possible that the constraints, including the assumed spatial structure and the assumed travellers' behaviour. point to a solution which is also contrary to this objective on a higher level. Finally, it may turn out that the proposed decision variables are not powerful enough to bring about the desired solution so that other instruments must be looked for.

To be able to answer this question, it is essential that the social objectives on a higher level are known. This is the reason why answering the question is a problem for society itself and not for a study such as the Dutch Integral Transportation Study.

If the question is answered positively, it is important to check once more whether all assumptions made correspond with reality. After that the results must be implemented and translated into a concrete work-planning and execution scheme.

If the answer to the question is 'no', it will be necessary to indicate which constraints, which decision variables and/or which objective function must then be used. A few possibilities present themselves. A single constraint can be changed, for example the spatial structure and, therefore, the given trip-matrix. The same optimization problem may then be solved. One can also try to include the minimization of the social costs for a given matrix of car-trips in a statement of the problem on a higher level.

It is obvious that this consciousness of objective function, decision variables and constraints must exist while the newly-stated optimization problem has next to be solved as well. If it is decided not to implement the results, all kinds of developments may take place in reality. If, for instance, a certain road is not constructed, changes will be made in the expected route choice and other roads will get more traffic and have to be expanded earlier; the expected choice between the transport modes may be affected as may the expected choice of destination; a number of trips will not take place or will occur at another point in time; finally, the planned land use pattern may not be realized. The consequences must be investigated and evaluated carefully in advance.

In this context we can continue the discussion in Chapter 8. The problem of defining the optimal spatial structure was reduced there to the minimization of the social costs for interurban transportation for two given matrices of car traffic and train transport. For every reduction consistency at least was required. The question we asked then was whether these reductions were allowable. We can consider this problem somewhat further now. At the beginning of this section, we stated that the results represent what will come about with unchanged travel behaviour and governmental policy. Is that representation accurate, taking into account the absence of consistency checks and feedbacks? And, if it is not so, how serious is that? Obviously the first question can be answered really well only after the application of the consistency checks and feedbacks. Still, even without these we may say something about it.

In the forecasting, it is generally assumed that the traveller makes his trip under travel circumstances comparable with those obtaining today. This generally means no congestion on highways. During the optimization process the trip-matrix is kept constant. Apart from fundamental changes in travel behaviour or in governmental intervention, this will in fact be true only if, on the network defined by the optimization process, the traveller again experiences circumstances in his journey comparable with those obtaining today. But in Section 10.3 it was stated that the optimal flow/capacity ratio is about 1,250 p.c.u. per lane in the average evening peak hour. Though this flow capacity ratio is rather higher than that experienced today, it cannot be spoken of as a fundamental change in travel circumstances. This implies that the representation given is a true reflection provided travel behaviour and governmental policy remain unchanged. Those parts of the network where the maximally-possible road infrastructure is insufficient, e.g. around the large towns, will prove exceptions to this. In the optimization we see there is a sharp increase of the objective function due to the high congestion costs; in reality there will be less traffic in those places. This means, however, that the social benefits of transport vanish. Because the benefits are not (as yet) taken into consideration, it seems very useful indeed to show the problems in those places fully.

Whether this discussion holds good for rail transport too may be questioned; but we will not go into that question here.

The question of whether the absence of feedbacks is serious can be answered negatively, at least for road traffic. In most places, the application of feedbacks would give the same representation. Furthermore in such places where the situation did change (around the large towns, for instance), feedbacks would only provide a—dangerous—veiling of the problems. For, on roads with serious congestion, the traffic will decrease and so will the traffic problems. This decrease in crowding and therefore in the social costs is immediately shown by the solution, while the decrease in social benefits caused by the omission of or change in certain trips is not shown. So feedbacks would have been misleading here and therefore are not desired.

The question of the seriousness of the absence of feedbacks must also be considered in relation to the answer to the question whether or not the solution is really desired. If the results are actually implemented, of course, greater accuracy is needed than when this is not the case. For the implementation of the solution in some places, a further specification will then be necessary for which the specific consistency can possibly be considered.

If what was said earlier in this section leads to a new statement of an optimization problem with another objective function and/or other decision variables and constraints, the feedbacks may, of course, become vitally important.

REFERENCES

Nederlands Economisch Instituut (1972). *Integrale Verkeers- en Vervoerstudie*, Staatsuitgeverij, 's-Gravenhage.

Steel, M. A. (1965). Capacity-restraint, a new technique. *Traffic Engineering and Control* (October).
Tweede nota over de Ruimtelijke Ordening in Nederland. (1966). Staatsuitgeverij, 's-Gravenhage.

List of Notation

A	matrix serving the network constraints or the network constraints and the given trip-matrix
A_m	network in which m investments are made (Section 4.3.3)
ARINV	coefficient for the investment cost in a link
ARINT	coefficient for the investment cost in a node
AMAINT	coefficient for the maintenance cost in a link
BRINV	coefficient for the investment cost in a link
BRINT	coefficient for the investment cost in a node
BMAINT	coefficient for the maintenance cost in a link
bn^{aj}	node on the path from a to j immediately preceding j (backnode)
C	vector of the dimensions of the links in a network
C^*	value of C which yields an optimal solution
C_t	dimensions of the links for period t
c_{ij}	dimension of link ij
c_{tij}	dimension of link ij for period t
$c_{ij}^{\min}$	minimal dimension of link ij
$c_{ij}^{\max}$	maximal dimension of link ij
c_a	dimension of the motor-road (Section 6.2.2)
CCR	coefficient in the travel time/flow relationship on a link
CCRE	coefficient (exponent) in the users' costs/flow relationship on a link
d_{ij}	length of link ij
d^{ab}	length of a path from a to b
d^{*ab}	length of the shortest path from a to b
$D1$	coefficient in the users' costs/flow relationship on a link (per year)
$D2$	constant term in the users' costs/flow relationship on a link (per year)
D1ACC, D1ACC′	coefficients in the costs of accidents/flow relationship on a link (per year and per p.c.u.)
D2ACC, D2ACC′	constant terms in the costs of accidents/flow relationship on a link (per year and per p.c.u.)
D1RUN, D1RUN′	coefficients in the vehicle operating costs/flow relationship on a link (per year and per p.c.u.)
D2RUN, D2RUN′	constant terms in the vehicle operating costs/flow relationship on a link (per year and per p.c.u.)
D1TIME, D3TIME	coefficients in the travel time costs/flow relationship on a link (per year)
D1TIME′	coefficient D1TIME without taking into account the fluctuations of the traffic flow

315

D2TIME	constant term in the travel time costs/flow relationship on a link (per year)
e_t	exploitation cost for the train (Section 6.2.2)
F	total (social) costs, objective function
F_{ij}	objective function on link ij
F_{ij}^{min}	minimal value for the objective function on link ij
F_{ij}'	derivative of F_{ij}^{min} (F_{ij} in Chapter 2) with respect to x_{ij}
F'^{ab}	minimal value for the derivative of the (minimal) objective function with respect to the flows on the related links on a path from a to b
F'^p	value for the derivative of the minimal objective function with respect to the flows on the related links on path p
F_{tij}'	derivative of F_{ij}^{min} with respect to x_{tij}
$F_t'^{ab}$	minimal value for the derivative of the minimal objective function with respect to the flows in period t on the related links on a path from a to b
F^u	upper bound for the objective function
F^l	lower bound for the objective function
F_i^l	lower bound for the objective function for a subset Q_i of the solutions
F_m	objective function for a network in which m investments are made (Section 4.3.3)
F^{ACC}	total costs of accidents
F^{RUN}	total vehicle-operating costs
F^{TIME}	total travel-time costs
f_t	frequency of the train (Section 6.2.2)
G	set of functions to describe the behaviour of the trip-makers
G_t	set of functions to describe the behaviour of the trip-makers in period t
g^{ab}	inverse demand function for the traffic from a to b
H	set of functions for the remaining constraints
H_t	set of functions for the remaining constraints in period t
I	total investments or total costs directly related to the transport network
I^0	fixed level of total investments
I_R^0	fixed level of investments for region R
I_t^0	fixed level of investments for period t
I_i^l	lower bound for the total investments for a subset Q_i of the solutions
i_{ij}	investments in link ij
i_{mij}	investments in link ij for travel mode m
i_{tij}	investments in link ij in period t
k	value of travel time
k_{auto}	value of time travelled by car
k_{train}	value of time travelled by train
KEXTRA	coefficient in the relationship for the value of travel time
KZERO	coefficient in the relationship for the value of travel time (value of time travelled on a road with no traffic on it)
L	set of links of the transport network
L_I	set of links in which an investment can be made
$\bar{L}_I$	set of links in which no investment can be made

l_{ij}	geographical length of link ij
l^{ab}	geographical length of the shortest path from a to b
$l^{*\,ab}$	distance from a to b as the crow flies
N	set of nodes of the transport network
N^D	set of destinations
N^I	set of intermediate nodes
N^O	set of origins
n_L	number of links of the transport network
n_{L_I}	number of links in which an investment can be made
$n_{\bar{L}_I}$	number of links in which no investment can be made
n_N	number of nodes of the transport network
n_P	number of transport relations
P	set of transport relations
Pa^{ab}	set of paths from a to b
Q	set of all feasible and infeasible solutions to the optimization problem
Q_i	subset of Q
Q_{ii}	subset of Q_i
S	total surplus, total consumers' surplus, total social surplus
S_a	total consumers' surplus of group a
S^{ab}	total consumers' surplus of travelling in relation ab
S_t	total surplus in period t
s	consumers' surplus per person
T	total generalized costs of travelling for the users
T^0	fixed total users' costs
T_a	total users' costs of group a
T_{ij}	total users' costs on link ij
t_{ij}	users' costs on link ij per tripmaker
t_e	final period
t_0	starting period
t^{ab}	minimum users' costs of travelling from a to b
t^{pab}	users' costs of path p from a to b
t^{mpab}	users' costs of path p from a to b by mode m
t_I^{ab}	minimum users' costs of travelling from a to b in situation I
TMAX	maximum value for the travel time per unit length
TMIN	minimum value for the travel time per unit length
TZERO	coefficient in the travel time/flow relationship on a link (travel time per unit length when there is no traffic on the road)
U	total social benefits
U^{ab}	total social benefits of travelling from a to b
U^{aa}	total social benefits of staying in a
U_t	total social benefits in period t
u	benefits per person
v_t	speed of the train
$\mathbf{X}$	vector of traffic flows and/or trip-matrix
$\mathbf{X}^*$	value of $\mathbf{X}$ that yields the optimal solution
$\mathbf{X}_t$	value of $\mathbf{X}$ in period t
x_{ij}	traffic flow on link ij (flowing from i to j)
x_{tij}	traffic flow on link ij in period t
x^a	number of trips originating or destinating in a

x^{ab}	number of trips from a to b
x^{0ab}	fixed number of trips from a to b
$x^{ab(n)}$	number of trips from a to b in the nth step
x_I^{ab}	number of trips from a to b in situation I
x_{ij}^{a}	number of trips originating or destinating in a, flowing on link ij
x_{ij}^{ab}	number of trips from a to b flowing on link ij
x_{ij}^{pab}	number of trips from a to b using path p flowing on link ij
$x_{ij}^{\alpha_n ab}$	number of trips from a to b on link ij after the α_nth part of the trip-matrix has been assigned to the network
Y	vector of Boolean variables, which indicate if investments are made
y_{ij}	Boolean variable, which indicates if an investment is made in link ij:
	$\quad y_{ij} = 0$ means no investment made in link ij
	$\quad y_{ij} = 1$ means an investment made in link ij
Z	total travel time costs
z	travel time per person
z_{ij}	travel time on link ij per person
ξ, η, ζ	coordinates
λ	Lagrange multiplier
π	discount rate

Author Index

Subject Index

323